# Sprinkler Hydraulics

Harold S. Wass Jr. · Russell P. Fleming P.E.

# Sprinkler Hydraulics

## A Guide to Fire System Hydraulic Calculations

Third Edition

Springer

Harold S. Wass Jr. (Deceased)
White Plains, NY, USA

Russell P. Fleming P.E.
Keene, NH, USA

ISBN 978-3-030-02594-6 ISBN 978-3-030-02595-3 (eBook)
https://doi.org/10.1007/978-3-030-02595-3



This Springer imprint is published by the registered company Springer Nature Switzerland AG
The registered company address is: Gewerbestrasse 11, 6330 Cham, Switzerland

# Preface to the Third Edition

It was an honor to be asked to update Harold Wass's book on sprinkler hydraulics for a third edition, and I have tried to tread as lightly as possible. I knew Harold, who passed away in January of 2000, the same year the second edition of this book was being published by the Society of Fire Protection Engineers. Harold was a 1950 graduate of MIT, and I knew him because he worked at Improved Risk Mutuals (IRM), based in White Plains, NY, not far from the Mount Kisco, NY headquarters of the National Automatic Sprinkler and Fire Control Association where I started my fire protection engineering career in 1975.

In 1980, following the MGM and Stouffers Inn fires that, respectively, took 84 and 26 lives, the Association was asked by CBS-TV to provide a demonstration of how fire sprinklers worked. IRM graciously allowed access to their sprinkler lab, and Harold Wass agreed to be interviewed by reporter Arnold Diaz and provide expert commentary for the event, for which I am forever grateful.

Wass's original *Sprinkler Hydraulics*, published in 1983, was based on rules in the 1980 edition of NFPA 13. The second edition, entitled *Sprinkler Hydraulics and What It's All About*, was published in 2000 but was only partially updated to the 1999 edition of NFPA 13, missing the consolidation of the storage standards into NFPA 13 and major reorganization of the document that took place with the 1999 edition. This third edition has been updated another twenty years to the 2019 edition of NFPA 13.

For the most part, fire protection hydraulic calculations are still being done in the same way as through all that time, using the Hazen–Williams method. Most of the necessary updates in the rules were motivated by attempts to clarify gray areas such as actual physical minimum areas of sprinkler coverage, projected sprinkler protection areas below sloped ceilings, and velocity limits. In such cases, we have retained Mr. Wass's commentaries on the background controversies while adding more recent code language that settled the arguments through the consensus process.

Fortunately, the rules for hydraulic calculations have not grown proportionately to the other rules of NFPA 13. The 2019 edition of NFPA 13 weighs nearly six times as much as the 1980 edition.

I reviewed both earlier editions of this book. The first review appeared in the Spring 1983 issue of the National Fire Sprinkler Association's *Sprinkler Quarterly*, the other in the Winter 2001 issue of SFPE's *Fire Protection Engineering*. In both cases, I praised Harold Wass's ability to take the mystery out of hydraulic calculations, and the conversational manner in which he was able to provide a context for the various rules within NFPA 13. I was surprised to find mention in the second edition that he took some exception to my statement that the book was written from an insurance authority's point of view. But I appreciated the fact that, after due soul-searching, he concluded that the insurance industry's bias is one of the more benign. All opinions expressed in this new edition are still those of Harold Wass; my role was only to update obsolete references.

I concluded my second review with the following remarks, and I stand by them: *Sprinkler Hydraulics* is Harold Wass's legacy, his gift to the fire protection community. It is one of a very few texts dealing with the subject of hydraulic calculations for sprinkler systems, and remains the best available.

Keene, New Hampshire, USA Russell P. Fleming P.E., FSFPE

# Second Edition Acknowledgments

This book is an update of the original book published in 1983 by my former employer, IRM Insurance. It would not have been possible without encouragement, support, and assistance from a number of people along the way. IRM was to be the publisher of this edition but the tides of change in the property insurance industry led to the dissolution of the organization a few years back. I am grateful to the Society of Fire Protection Engineers for assuming the role of publisher.

IRM Insurance was a small property insurance organization that was unique in terms of its dedication to the principles and practices of fire protection. There were a number of very fine people I had the pleasure of working with and learning from. I would like to dedicate this book to the spirit of IRM and its people, from Bud Bolz and his successor, Bruce Jamieson, on through the entire organization, who sustained the special environment that made this book possible.

Very particular thanks are due to Ike Siskind who, as Vice President of Engineering at IRM, spurred me to expand my in-house manual into the original book and has continued his support of the project to this day. In the production of this manuscript, Carol Shackelford wrestled with the text, including all of the numbers and messy equations. Her unfailing dedication to the lengthy and tedious editing process was indispensable.

Kevin Kimmel, when Vice President of FPE Software, Inc., was gracious enough to provide input for the section on "Personal Computer Programs for Hydraulic Calculations." His extensive knowledge and first-hand experience filled a void.

I also want to thank my wife, Nancy, who will never read this book but did not consider my idiosyncratic use of time to produce it grounds for divorce.

Harold S. Wass Jr.
2000

# Contents

# Introduction

It was back in 1874 that Henry Parmelee patented his first automatic sprinkler. Installation standards evolved, with limitations on the distance between sprinklers and branch lines and the number of heads that could be supplied by a given pipe size. Significant changes in the so-called pipe schedule were made in 1885, 1905, and 1940. The standard sprinkler itself gradually evolved, with a major change being the introduction of the "spray" sprinkler in 1953, and the development of fast response sprinklers in the 1980s. By the dawn of the twenty-first century, the various fast response sprinklers constituted the majority of the marketplace.

Over the first hundred years, automatic sprinklers installed on a pipe schedule demonstrated a high degree of reliability in controlling fires. Today, however, the pipe schedule system is receding into history with almost all new installations being "hydraulically designed." There are several reasons for this.

In the late sixties and early seventies, the problem of protecting modern warehouses with high storage configurations, and the proliferation of plastics, received intensive study. It was found that sprinkler discharge rates in excess of those obtained from pipe schedule systems were usually needed. Thus, systems need to be hydraulically calculated to determine the pipe sizes required to deliver the higher rates of discharge. Coincidentally, there was the advent of powerful hand calculators, soon followed by personal computers, at an affordable price. Even the most complex hydraulic calculations could be accomplished economically. In this environment, standards for hydraulically designed systems in ordinary and light hazard occupancies first appeared in the Sprinkler Standard, NFPA 13, in 1972.

It was soon recognized that hydraulically designed systems were, in most cases, more economical to install than the traditional pipe schedule systems and most sprinkler contractors soon acquired the hardware and software that could easily make the calculations.

The 1991 Edition of NFPA 13, for the first time, placed severe restrictions on the use of pipe schedule systems.

The extra hazard pipe schedule was relegated to the Appendix for reference on existing systems. All new extra hazard systems were required to be hydraulically



H. S. Wass Jr. and R. P. Fleming P.E., *Sprinkler Hydraulics*,
https://doi.org/10.1007/978-3-030-02595-3_1

calculated. The use of the pipe schedule for light and ordinary hazard occupancies was severely restricted. For new installations exceeding 5,000 sq. ft., the pipe schedule may now be used only if there is a minimum residual pressure of 50 psi available at the elevation of the highest sprinkler. With most water supplies, a pipe schedule system would generally

deliver a higher density than required for the minimum design area of 1,500 sq. ft., when hydraulically calculated. The committee responsible for NFPA 13, however, was concerned with the growing use of pipe schedule systems with a low-pressure water supply, where larger pipe sizes would be required if the hydraulic design requirements were to be met. This is a valid concern although it could be argued (and was) that the 50-psi minimum residual pressure is unduly restrictive.

Problems usually go hand-in-hand with change. Unfortunately, we suspect that some of the problems associated with hydraulically designed systems will always be with us. We will attempt to explain the "hows" and "whys" of hydraulically designed systems, along with a bit of the underlying theory, and discuss the associated problems, as we perceive them.

The most significant change in recent years has been the development of a variety of sprinklers with specific characteristics designed to increase their effectiveness in specific fire scenarios. The residential sprinkler and the Early Suppression Fast Response (ESFR) sprinkler are two of the most prominent examples. We will take a brief look at these developments and discuss yet another set of problems that these sprinklers introduce.

We do not mean to be negative. While the sprinkler industry is a special interest group that may not always be totally objective, the end result of many of their efforts is commendable. Building codes are giving increasing recognition to the important contribution sprinklers can make to life safety. There is greater recognition of the need for sprinklers in buildings. A number of dramatic fires have demonstrated the limitations of even the best fire department when a properly designed automatic sprinkler system is not in place.

General awareness of sprinkler systems may be increasing. This was illustrated in a subtle way by a news story on the lawsuit challenging the all-male tradition at The Citadel. The story, appearing in the May 29, 1994 issue of The New York Times, included the following sentence:

"The courtroom proceedings, with all the drama of an automatic sprinkler system, passed tediously back and forth over the same ground."

Some of us would argue that an automatic sprinkler system, when doing its job, can be very dramatic but, of course, the writer was thinking of the normal unobtrusive (except to interior decorators) mode.

Rolf Jensen, a very respected name in fire protection engineering, wrote a chapter entitled *The Historic Development of the Sprinkler Standard,* for the 1994 *Automatic Sprinkler Systems Handbook* published by the National Fire Protection Association. Referring to the many changes in NFPA 13 over the years, he made the following observation:

"Each of these changes shows how the driving forces of sprinkler technology evolved and are still evolving. They have been and will continue to be political and economic in nature, but a true technology has begun to take over."

As he made clear, NFPA 13 is the product of people. People are imperfect, but my first and secondhand knowledge of the people gives me reasonable confidence in the end product.

In the early 1960s, I recall a consultant whose name is still around, though he is not, privately accusing the NFPA of being controlled by the "sprinkler interests." Rolf Jensen, in the article mentioned above, pointed out that "there has been a progressive evolution to a uniform balance of interests." He referenced a table of "Voting Membership Interests." Most notably, this table showed that in 1960, 16 of the 27 members of NFPA 13 were insurance people, whereas in 1989, only 4 of 28 represented the insurance industry. Having spent my past 30 years on the insurance side of the fence, I may be less enthusiastic than he is about the "progressive evolution."

I recall Russ Fleming's review of my original Sprinkler *Hydraulics.*

While generally favorable, it included a caveat to the effect that it was written from an insurance company's point of view. This gave me pause since, of course, words like "objective" floated around in my self-image. To some degree, everyone has his/her own acquired perspective and axe to grind. In its defense, the insurance industry is concerned about the bottom-line performance of sprinklers since it pays for the losses. The sprinkler industry representatives are trying to sell their products. Consultants, whose representation on NFPA 13 went from 0 to 6 between 1960 and 1989, could logically, if subconsciously, have a bias toward complexity.

# Automatic Sprinkler Systems—A Brief Overview

Anyone wishing to gain a full understanding of sprinkler systems should recognize that there are other important subjects such as types of sprinklers, other hardware, layout requirements, reliability, and sprinkler demand, along with related concerns, such as alarms and minimizing water damage, that are beyond the focus of this book.

A sprinkler system should be viewed as just that, a "system" composed of the following main elements:

1. Single or multiple water supplies.
2. Piping, underground and overhead, connecting the water supply, or supplies, to the sprinklers.
3. Sprinklers.
4. Associated hardware, such as control valves, check valves, dry pipe valves, and fire department connections.
5. Alarms.

For the first hundred years after the introduction of the Parmelee sprinkler, few things changed as little as the automatic sprinkler system. As we will be discussing briefly, sprinkler technology has made major strides in the past few decades. Nevertheless, compared to much of the world around us, the changes have been fairly undramatic. Despite some sophistication in the design of sprinklers, we are still using simple mechanical devices.

It is fair to say that a sprinkler system is not only rather unsophisticated, but it is inefficient. Only a very small percentage of the sprinklers installed will ever discharge water. When water flows, it may not be in response to a fire. Accidental water flows, "sprinkler leakage" as it is called in the insurance business, from such causes as mechanical damage or frozen pipes, are fairly common. When sprinklers operate in a fire, significant water damage frequently occurs.

Is there a better way? Alternative extinguishing agents such as carbon dioxide and clean agents are available. Thanks to modern electronics, we can endow mechanical devices with all manner of "intelligence." With space age technology,



H. S. Wass Jr. and R. P. Fleming P.E., *Sprinkler Hydraulics*,
https://doi.org/10.1007/978-3-030-02595-3_2

there may be a "better way," but only if you ignore two critical considerations—cost and reliability. The lack of sophistication in the sprinkler systems may, on balance, be a virtue. There may not yet be a better way.

Typically, in fact, sprinkler systems are unfairly maligned, particularly by the ill-informed media. The critic who points out that the water damage from the sprinklers exceeded the fire damage fails to consider what the fire damage would have been without the sprinklers, not to mention the water damage from fire department hose streams. The spokesman for a hotel that has sustained a multiple-fatality fire defends the lack of sprinklers by telling the press that sprinklers are for property protection, not protection of lives, endowing sprinklers with a mysterious ability to discriminate between life safety and property protection. The politician defending the deficient safety code for nursing homes is quick to emphasize that the people in the nursing home died from the products of combustion, not from the flames to which a sprinkler system reacts, as if the one were not related to the other.

The subject of sprinklers as they relate to life safety is complex and beyond our scope. It should be emphasized that sprinklers, alone, do not solve a life safety problem, although the residential sprinkler is a step in that direction. On the other hand, we do believe that sprinklers are an important part of the solution in most buildings. Proper exits, vertical and horizontal compartmentation, control of interior finishes, control of the amount and nature of the combustible contents, and smoke control systems are pertinent but difficult to police and easy to subvert. A sprinkler system, if properly designed, installed, maintained, and supervised, may be the vital ingredient when all else fails.

# NFPA 13

The National Fire Protection Association standard number 13, the Standard for the Installation of Sprinkler Systems, is universally recognized in the United States as the source of the minimum standards for automatic sprinkler systems.

The NFPA is an independent, non-profit organization. NFPA standards are sometimes referred to as “consensus” standards. The NFPA committees responsible for writing the standards are made up of a diverse group of people knowledgeable in the subject matter. The committee responsible for NFPA 13, for example, is composed principally of representatives of sprinkler system hardware manufacturers and installers, insurance company representatives, enforcement authorities, and consultants. What might be considered special interests abound on NFPA committees, but the NFPA is careful to ensure a balance of interests. Further, there is a public review and comment procedure for all changes proposed by the committee, followed by a vote of the NFPA membership, and a final issuance by a standards council following the possibility of appeals.

We have the utmost respect for the NFPA committee process, and the caliber of the people typically found on these committees. Opinions and comments expressed in this book should be read with this in mind.

It should be recognized that an NFPA standard cannot, and should not, do more than set out basic guidelines. Further, since these standards, and NFPA 13 in particular, are widely used as the basis of governmental regulation, it is critical that all of the provisions within the standard are reasonable, practical, and defensible, based on current knowledge and technology. With constantly changing knowledge and technology, it is unavoidable that there is a time lag in the codes. Typically, these codes are revised on three-year cycles, which is reasonable considering the nature of the process.

NFPA rules do provide for the issuance of what they call a Tentative Interim Amendment to deal with an issue that is deemed to be of an emergency nature that surfaces between editions of the standard.



H. S. Wass Jr. and R. P. Fleming P.E., *Sprinkler Hydraulics*,
https://doi.org/10.1007/978-3-030-02595-3_3

All references to, and quotations from, NFPA 13 in this book refer to the 2019 Edition unless otherwise indicated.

Many readers of this book will be people who keep the current edition of NFPA 13 at their fingertips. If that does not include you, we strongly recommend that you obtain the current edition. Better yet, buy the *Automatic Sprinkler Systems Handbook*. This contains the complete text of NFPA 13 interlaced with commentary, figures, and examples. In their words, "it explains the philosophy and history behind requirements…so you gain a full understanding of why and how to apply provisions correctly." These publications can be obtained from the National Fire Protection Association, 1 Batterymarch Park, P.O. Box 9101, Quincy, MA 02269–9101, for a moderate price. All NFPA codes and standards can also be viewed freely online at www.nfpa.org.

You may now be wondering why you wasted your money on this book. This book focuses on hydraulic design and attempts to cover the subject in much greater depth. If you must set priorities for procuring publications on the subject of automatic sprinklers, NFPA 13 or the Handbook should be Number 1. We will argue for Number 2, but consider the source.

Before we leave this subject, one more unsolicited and unpaid plug for an NFPA publication. A standard was first published in 1992 as NFPA 25, *Inspection, Testing and Maintenance of Water-Based Fire Protection Systems.* However well designed and well installed the sprinkler system, proper inspection, testing, and maintenance are critical to its long-term reliability and NFP A 25 provides guidance in the form of an enforceable standard.

# Those Magic Words...Hydraulically Calculated

"Hydraulically calculated" has a nice ring to it. Really scientific stuff.

It must be good.

The first thing that must be understood is this:

There is nothing inherently desirable about a calculated sprinkler system.

Many people take exception to this statement, but they must agree that a calculated system, more properly referred to as a hydraulically designed sprinkler system, may provide excellent protection or it may provide poor protection. A hydraulically designed system must always be evaluated in terms of

1. Sprinkler and hose stream demand.
2. System design.
3. Available water supplies.

It should be understood that all hydraulic calculations are approximations. The Hazen–Williams formula, which is the accepted basis for fire protection hydraulic calculations, is empirically derived; it is not a scientific truth. One of the elements of the formula, the "C" value, is only a guess. The calculation routine, particularly with grids, involves further approximations. Do not be misled when you read that 923.7 gpm at 71.66 psi is required. At best, everything to the right of the decimal point is meaningless. It should be acknowledged, however, that the hydraulic calculations do have a greater degree of precision than the other two elements that must be evaluated—sprinkler demand and water supplies.

For better or for worse, the pipe schedule systems are a vanishing breed.

NFPA 13 formally moved in this direction for the first time in the 1991 Edition when it added the following wording:

> 5–2.2.1...The pipe schedule method shall be permitted only for new installations of 5000 sq. ft. or less or for additions or modifications to existing pipe schedule systems.

This met with some opposition during the public comment process for this revision. The committee responded by adding an "exception":



H. S. Wass Jr. and R. P. Fleming P.E., *Sprinkler Hydraulics*,
https://doi.org/10.1007/978-3-030-02595-3_4

> Exception No. 1: The pipe schedule design method shall be permitted for use in systems exceeding 5000 sq. ft. when the flows required in Table 5–2.2 are available at a minimum residual pressure of 50 psi at the elevation of the highest sprinkler.

Table 5–2.2 established the "Water Supply Requirements for Pipe Schedule Systems." For Ordinary Hazard, the "Acceptable Flow at Base of Riser" was 850–1500 gpm with a residual pressure at the elevation of the highest sprinkler of 20 psi. It was further stated that "the lower flow figure shall be permitted only where the building is of noncombustible construction or the potential areas of fire are limited by building size or compartmentation such that no open areas exceed 4000 sq. ft. for Ordinary Hazard."

The published committee rationale for this restriction on pipe schedule systems:

> The committee believes that hydraulically designed systems are superior to systems designed with the schedule method. In addition, some designers have arbitrarily selected the pipe schedule method when the systems could not be designed hydraulically due to limitations on pressure.

The "committee", of course, is not a monolithic entity and the extent of their belief in the superiority of hydraulically designed systems is an open question. It is a matter of record that the NFPA 13 representative from Factory Mutual expressed opposition to these restrictions on pipe schedule systems. And it certainly was not because Factory Mutual was uncomfortable with hydraulic design!

Prior to the NFPA mandate, most new ordinary hazard systems were hydraulically designed. Why? They were cheaper; that is, they could be designed with a smaller pipe. Except when bludgeoned by an insurance carrier to provide a cushion, the systems are designed right to the water supply. The old-fashioned pipe schedule system, if calculated, will typically provide a substantial cushion. This can be particularly helpful in a retail occupancy just before Christmas or in a light occupancy with some temporary storage while undergoing renovations.

The main concern of NFPA 13, presumably, was with areas such as Chicago that have low water pressure. I have never seen anyone suggest, let alone provide evidence, that the historical sprinkler performance in such areas was inferior. On the other hand, let us look at a system we just pulled out of the air on the first try.

The building is 100 ft. wide and 144 ft. deep with an Ordinary Hazard Group 2 occupancy. A density of 0.20 gpm per sq. ft. over the most remote 1500 sq. ft. is needed. Suppose we install a pipe schedule system, center feed with 5 sprinklers on each side of the cross main. The sprinkler spacing along branch lines is 10 feet and the branch line spacing is 12 feet, with 12 pairs of branch lines. It is a one-story building with the sprinkler deflectors 12 feet above the floor. We will assume a wood roof and NFP A 13 states that the "acceptable flow" is 1500 gpm. Now just what does the required "minimum residual pressure of 50 psi at the elevation of the highest sprinkler" mean? It is easy to determine the residual pressure at the base of the riser with 1500 gpm flowing if a hydrant flow test is performed. The residual pressure as you go out into the sprinkler system drops as the elevation increases and drops because of friction loss in the piping. It would be absurd, however, to carry a flow of 1500 gpm through the system. Therefore, we will assume that what they really mean is a residual pressure at the base of the riser of at least 50 psi at a flow of 1500 gpm, with the 50 psi minimum increased by the elevation of the highest sprinkler, or in this case by about 5 psi, meaning that we must have about 55 psi available at the base of the riser at a flow of 1500 gpm. (If you are new to the field and do not understand all of this, you might wish to move on and return later.)

If we assume a water supply with a static pressure of 70 psi and a residual pressure of 55 psi at a flow of 1500 gpm, we find that the pipe schedule system set forth above will deliver a density of about 0.21 gpm per sq. ft. over the most remote 1500 sq. ft. If we had a very strong water supply (in terms of volume available) and the static pressure was 60 psi, rather than 70, the system would only deliver about 0.19 over 1500 sq. ft. Perhaps the folks on NFPA 13 know what they are doing!

It gets worse if you go to the other end of "area/density curve." 1500 sq. ft. is the minimum, and normally used, area. The high end of the curve is 0.15 gpm per sq. ft. over 4000 sq. ft. Using the above two water supplies, the system will deliver only 0.14 and 0.15, respectively.

While I have not presented the worst case, the five sprinkler branch lines present one kind of worst case in that the hydraulics would work out more favorably with either more or fewer sprinklers on the branch lines. Why is that? I will let you figure that out. Hint: Refer to Appendix F for the rules for sizing pipe. While I selected a "worst case" width for the building, I did not select a "worst case" depth since the deeper the building, the longer the cross main and the greater the friction loss.

Perhaps the most serious criticism of the good old pipe schedule is the absence of special provisions for side feed systems. For example, if we cut the above building in half, making it 50 feet wide instead of 100, and it was a side feed system (perhaps because of a slight pitch in the roof), in the first example where the system delivered 0.21 gpm per sq. ft., the density would drop to 0.17 gpm per sq. ft. even though the branch lines in the two systems are identical. The reason is very simple. Each section of cross main is supplying half the number of sprinklers and it is the number of sprinklers being supplied that governs the pipe size. The *flowing* sprinklers, however, are the same in both cases.

# A Word About the Math

Although not indispensable, a hand calculator with the square root function is highly desirable for even the simplest hydraulic calculations. A calculator with scientific notation is very helpful for calculating velocity pressure. For anyone not wishing to be totally dependent upon tables for friction loss, a calculator with the $y^x$ function is needed A programmable calculator can be programmed to do many things, depending upon its storage capacity and your programming ability. A personal computer is only limited by the software with which you equip it.

Some of the useful constants appearing in this book are in scientific notation. For those of you not mathematically inclined, a word of explanation is in order. Scientific notation is a convenient way of handling numbers that are very large or very small. It consists of a number, the mantissa, multiplied by 10 raised to a power. A simple example: 231 is expressed in scientific notation as $2.31 \times 10^2$. The number, 2.31, is the mantissa, which multiplied by $10^2$, or 100, equals 231. For those not at home with exponents, $10^0 = 1$, $10^{-1} = 0.1$, $10^{-2} = 0.01$, etc. For example,

$$4.86 \times 10^{-4} = 0.000486.$$

All friction-loss equations involve exponents. Since occasionally we manipulate these exponential equations, a brief review of exponents may be helpful. For example,

$$\begin{aligned} 25^0 &= 1 \\ 25^1 &= 25 \\ 25^{1.85} &= 385.6 \\ 25^2 &= 625 \end{aligned}$$

Calculating $25^{1.85}$ is an example of where the "$y^x$" function on a calculator is useful. In the absence of this function, it is necessary to use a logarithm table or a table of numbers raised to the 1.85 power.



H. S. Wass Jr. and R. P. Fleming P.E., *Sprinkler Hydraulics*,
https://doi.org/10.1007/978-3-030-02595-3_5

Consider this

$$\text{If } a = b^{1.85}$$

$$\text{Then } b = a^{0.54}$$

This is determined by dividing the exponents on each side of the equation by 1.85. The implicit exponent of "a" in the first equation is "1," which divided by 1.85 is approximately 0.54. This explains the 0.54 exponent, which occasionally appears in this book.

Actually, the reciprocal of 1.85 is a repeating decimal, 0.540540540…but 0.54 appears to be sufficiently accurate. If you have a calculator with the "$y^x$" function, however, try the following:

Enter 1000 and raise it to the 1.85 power. Now calculate the 0.54 power of the answer. This should bring you back to about 1000, but my very reputable calculator comes up with about 993.1. Repeat this procedure, using the repeating decimal as far as the display will permit in place of simply "0.54" and you should get an answer very close to 1000. The number 0.54 is sufficiently accurate for most purposes, but there are a few instances where puzzling discrepancies can arise unless you are aware of the possible consequences of this approximation.

A square root is normally indicated by the radical sign, such as in the sprinkler flow equation, $Q = k\sqrt{p}$. However, this could also be expressed as $Q = kp^{0.5}$ or $Q = kp^{1/2}$.

# A Few Words About the Units of Measurement

The United States is the only remaining highly industrialized nation still using the English system of measurement. In the mid-1970s, there was a movement toward adopting the metric system, but little has happened since. The resistance is quite natural. When you work with an unfamiliar set of units, you no longer have a sense of the numbers. You have to develop that sense all over again. But international trade considerations lend some urgency for us to fall in line. Federal legislation, with that in mind, is forcing the federal agencies to move to metric. The General Services Administration, which is responsible for building all federal buildings, is slowly following the metric mandate. The associated sprinkler·installations will bring this home to the United States fire sprinkler industry.

The first confusion for many Americans is the reference to the International System of Units which is commonly referred to as the SI system. It turns out that "SI" comes from the name in French, Systeme Internationale d'Unites. These units were adopted by an international body, the 11th General Conference on Weights and Measures, meeting in Paris in 1960. It is appropriate that the meeting was in Paris because France was the originator of the metric system after the French revolution, with the formal adoption of the system in 1799.

The users of NFPA 13 saw the introduction of metric equivalents in the 1978 Edition, those mildly annoying parenthetical numbers we are so familiar with. This also provided an opportunity for more errors to creep into the standard. Prior to the 1991 Edition, below the density/area graphs appeared the following: For SI units: 1 sq. ft. = 0.0920 $m^2$. The 1991 Edition correctly shows 0.0929 $m^2$ as the equivalent.



H. S. Wass Jr. and R. P. Fleming P.E., *Sprinkler Hydraulics*,
https://doi.org/10.1007/978-3-030-02595-3_6

The International System of Units (SI) consists of seven base units:

| | |
|---|---|
| Length | meter (m) |
| Mass | kilogram (kg) |
| Time | second (s) |
| Electric current | ampere (A) |
| Temperature | kelvin (K) |
| Luminous intensity | candela (cd) |
| Amount of substance | mole (mol) |

Other units are derived from these. Four of these basic units relate to sprinklers: length, mass, time, and temperature.

Length. An old (1944) physics book of mine has the following to say on the subject:

The standard meter is a bar of platinum–iridium of X-shaped cross section. The meter is defined as the distance between two fine transverse lines engraved on this bar. For many years, the foot was defined as one-third of the distance between two lines on a similar one-yard standard; but to avoid the necessity of maintaining two standards of length when one is sufficient, the United States yard is now defined by the relation

$$1\,\text{yard} = \frac{3600}{3937}\,\text{m}\,(\text{exactly})$$

Simple arithmetic indicates that a foot is approximately 0.3048 m.

Mass. Mass is a quantitative measurement of inertia, an invariant property of a body. Weight is the force required to support the mass when it is at rest. Thus weight depends upon the acceleration of gravity, which varies slightly over the surface of the earth, and varies considerably when we move out into space. In the American engineering system, we ask that weight, or force, be essentially the same at the earth's surface as the mass, and both are expressed in pounds even though they are not the same units. The force exerted by a body varies slightly at different places on the earth because of the variations in gravitational attraction. Thus, the force exerted on a body in Boston exceeds the force in the "mile high city" of Denver by a factor of about 1.0008. Not to worry. Whereas a standard pound originally related to sea level at a 45° latitude, along the way it was simply stipulated that the mass of a pound shall equal 0.4535924277 kg.

Time. For some strange reason, everybody is in agreement despite the awkward 60 s/min, etc.

Temperature. The Kelvin measurement is the more common Celsius temperature plus 273.15°. For sprinkler systems, the Celsius temperature is appropriate. On the Celsius scale, 0° is the freezing point of water and 100° is the boiling point at normal atmospheric pressure (14.7 psi). At one time, the Celsius scale was more commonly known as the centigrade scale, reflecting the 100° interval between the base points. More recently, Celsius has come into common use, honoring the inventor, Anders Celsius, an eighteenth-century Swedish astronomer.

To convert from Fahrenheit to Celsius:

$$\text{degrees C} = 5/9(\text{degrees F} - 32)$$

Let us look at the English units that are commonly used in the sprinkler trade:

Feet
Square feet
Pounds per square inch (pressure)
Degrees (temperature)
Gallons
Gallons per minute
Gallons per minute per square foot (density)

Two of them, feet and degrees, are base units that have already been discussed. We will look at the other units, which are derived.

Square feet. Since a foot is 0.3048 m, a square foot is obviously 0.3048 squared, or about 0.0929 square meters ($m^2$). Conversely, a square meter is approximately 10.764 sq. ft.

Pounds per square inch (psi). The SI unit of force is (remember the earlier discussion of mass vs. force) the newton. The newton is defined as the force that imparts an acceleration of one meter per second, per second, to a mass of one kilogram. (Newton's second law: Force = Mass x Acceleration).

We have that "acceleration" number here, the "g" which we have discussed elsewhere. It depends upon where you are. There may be no consistency on this but NFPA, in its *Manual of Style*, specifies that one pound equals 4.448 newtons, making the newton equal to a force of 0.225 lb. The SI unit of pressure is the pascal (named for Blaise Pascal, a seventeenth-century mathematician, physicist, and religious philosopher). A pascal is one newton of force per square meter. One inch equals 0.0254 meters. Squaring that, a square inch is about 0.00064516 $m^2$, or a square meter is about 1/0.00064516, or about 1550 sq. in.; Multiplying 1550 by 4.448 yields 6894.4, very close to the NFPA conversion factor: 1 psi = 6.895 kilopascals. In Europe, however, it is common to use the bar, which is 100 kPa, and NFP A 13 uses this, 1 psi = 0.0689 bar, for its metric equivalent.

Gallons. A liter is 1/1000 of a cubic meter. One gallon is 3.785 L.

Gallons per minute. Since the units of time are common, one gallon per minute is equal to 3.785 L per minute.

Gallons per minute per square foot. Using 3.785 to convert from gallons per minute to liters per minute and dividing by 0.0929 to convert from sq. ft. to sq. meters yields 40.743, which is very close to the "official" NFPA conversion number of 40.746. The 1999 Edition of NFPA 13 was the first to simplify (L/min)/$m^2$ to mm/min, such that 1 gpm/$ft^2$ = 40.746 mm/min.

It should be noted that, to make NFPA 13 more user friendly in regions outside the United States, a significant change was made in the 2016 Edition of the standard. It moved from using exact metric conversions to soft or approximate conversions, eliminating the irregularity of the metric equivalents.

Throughout this book, SI units are conveniently ignored and we apologize to those of you who work in those units. The numbers get pretty messy as it is.

See Appendix D for a summary of conversion factors between U.S. and SI units.

# The Evolution of the Sprinkler: Choose Your Weapons with Care

What NFPA 13 refers to as the "Old-Style/Conventional Sprinkler" was superseded in 1953 by the spray sprinkler, which today is still considered the "standard sprinkler." While some small benefit may be derived from replacing the old-style sprinklers, many are still in service although sample testing of sprinklers over 50 years old is required by NFPA 25. Some sprinklers of World War I vintage contain solder alloys that are suspect with regard to long-term integrity, which led to an NFPA 25 requirement to replace all sprinklers manufactured prior to 1920. The old-style sprinkler discharged about half of the water upward, generally within a radius of only a few feet. (UL 199, the Underwriters Laboratories Standard for Automatic Sprinklers, says "approximately 40%" upward.) The water thrown upward then drips from the ceiling in relatively large drops. The pattern of distribution from the old-style sprinkler is less uniform and the droplet size less favorable.

The spray sprinkler discharges all of its water downward in a parabolic pattern. The reasoning behind this was that most fires originate below the sprinklers; therefore, directing all of the water in that direction should prevent horizontal fire spread while the fine water droplets cool the ceiling gases sufficiently to prevent ignition of the roof. This is generally a valid assumption although we have all read of roof fires not controlled because they were "above the sprinklers." For whatever reasons, the Europeans continue to prefer a design similar, but not identical to the old-style sprinklers which they call the "conventional style sprinkler."

In the early nineties, 12 large-scale fire tests were conducted by Factory Mutual Research Corporation to compare the relative performance of European conventional style sprinklers and the U.S. spray sprinklers.[1] They used two scenarios. One was rack storage of the standard plastic commodity, polystyrene cups in cartons, ten feet high in a 30-foot building. The other scenario was rack storage of a Class II

[1]" Comparison of European Conventional and U.S. Spray Sprinklers" by Bennie G. Vincent and Hsiang-Cheng Kung, Factory Mutual Research Corporation, Journal of Fire Protection Engineering, 1993.



H. S. Wass Jr. and R. P. Fleming P.E., *Sprinkler Hydraulics*,
https://doi.org/10.1007/978-3-030-02595-3_7

commodity, double-walled cartons with a sheet metal liner inside, 20 feet high in a 30-foot building.

The ignition location was varied from directly below a sprinkler to between sprinklers. One-half inch (k = 5.6) and 17/32 inch (k = 8) diameter orifice sprinklers were used, all rated at 165 °F. The 12 tests meant there were actually six comparison tests. The spray sprinklers performed better in four of the six. One of the two instances where the European sprinkler outperformed the spray sprinkler was with the Class II commodity where the ignition was directly below a sprinkler. This was attributed to the localized high density directly below the sprinkler from the water directed upward, then dripping down. Generally, more conventional sprinklers opened than spray sprinklers. No sweeping conclusions could be drawn from these limited tests, which did not replicate the vast majority of real-life fire scenarios. They did, however, tend to support the view that the spray sprinkler will outperform the European conventional style sprinkler in most cases. And, interestingly, while the Europeans were particularly concerned about the spray sprinkler with high piled storage, the spray sprinkler had the biggest edge with the high challenge plastic fires. Of course, since these tests were conducted by the folks who originally developed the spray sprinkler, our European friends may have questioned their objectivity.

During the past 40 years or so, the evolution of sprinkler protection has accelerated greatly. In the late 1970s, much attention was focused on residential sprinkler protection, recognizing that many lives are lost in residential fires. Historically, sprinklers had been designed for property protection with life safety simply a byproduct. The primary goal for residential systems, however, was life safety. When residential sprinkler tests were initiated, it soon became clear that the conventional sprinklers did not always operate before the room environment became life threatening. A faster responding sprinkler was needed.

A sprinkler does not activate immediately upon being exposed to the temperature at which it is rated. There is thermal inertia, that is, the thermal element does not instantaneously assume the ambient temperature. The delay can be significant. This has never been considered to be all bad. While fast actuation of the sprinkler closest to the fire is obviously desirable, sprinklers remote from the fire that tends to open more quickly can reduce the flow from the sprinklers over the fire.

The first known measurements of "sensitivity", or actuation time, were, interestingly, made in 1884 when the sprinkler was in its infancy. A wide range of actuation times were noted. (Modern spray sprinklers still exhibit a wide range.) Grinnell introduced the "Duraspeed" sprinkler in 1935 with the claim that it was much faster, implying that faster was better although nobody could be sure. The traditional Underwriters Laboratories "Response Test" merely sets an upper limit on sensitivity although nobody was entirely sure what that upper limit should be.

The first so-called "quick-response" sprinkler, and so listed by Underwriters Laboratories, for whatever that meant, once again came from Grinnell. It was the Grinnell Model F931 employing an "electronic squib." It was ahead of its time and eventually discontinued.

Underwriters Laboratories developed a standard, UL 1626, for testing residential sprinklers. As part of a residential sprinkler development program funded largely by the U.S. Fire Administration, the standard focused on the special attributes required to be effective in the residential environment with limited water supplies, fast response, and high wall-wetting capabilities.

At about the same time as the residential sprinkler was being developed, development was underway on what was called the "large-drop" sprinkler, heavier fire power for high challenge storage configurations. The large-drop sprinkler has a large (nominal 0.64 in. diameter with a nominal k of 11.2) orifice and, as the name implies, was designed to deliver relatively large droplets better able to penetrate the fire plume. This should not be confused with the conventional extra large orifice sprinkler, which has the same orifice size and "k" factor. The large-drop sprinkler, now considered a type of "Control Mode Specific Application" sprinkler, has a different deflector design and the sprinklers are not interchangeable.

For the first time, a minimum pressure and a number of "design sprinklers" were specified rather than density and area. In most cases, a minimum sprinkler pressure of 25 psi was specified although higher pressures were required in a few scenarios. The number of design sprinklers varied from 15 to 30 for wet systems, with a larger number of design sprinklers where dry systems are permitted. The protection area per sprinkler is limited to 130 sq. ft. except it was 100 sq. ft. for what was defined as combustible obstructed construction. Minimum spacing was 80 sq. ft.

A protected area could be determined from the sprinkler spacing and the number of design sprinklers. This area was then subject to the "1.2 times the square root of the area" rule (discussed elsewhere) used for conventional "density/area" designs.

Calculating a large-drop system was really no different from calculating a conventional system. In both cases, you determined the minimum discharge per sprinkler, but in the case of the large-drop system, the minimum discharge was determined by the minimum pressure required.

Returning to the residential sprinkler, the 1980 Edition of NFPA 13D specified the use of listed residential sprinklers even though they did not exist. Again, Grinnell came through with the first UL listed residential sprinkler in June 1981, the first of a new generation of fast-response sprinklers. Fast-response sprinklers, quite obviously, have less thermal inertia. A fusible link type, for example, has a much thinner fusible link and the glass bulb type has a more slender cylinder.

We have been talking about the response time of sprinklers. If you are concerned about response time, the next question is how do you measure it?

As you might guess, Factory Mutual came to the rescue when in 1981 they introduced the concept of Response Time Index, commonly referred to as RTI.

Convected heat, heated air rising to the ceiling, is the primary means by which a sprinkler is actuated. Radiated heat generally plays only a very minor role. In this context, a "plunge test" was developed to measure the sensitivity of the heat-sensing element. This involved a constant- temperature, constant-velocity air stream passing through a duct. The sprinkler, at room temperature, was plunged into the air stream. A measurement was made of the time, in seconds, required to raise the temperature of the heat-sensing element to about 63% (1–1/e, to be exact) of the temperature of the heated air stream. This measurement was called the "tau factor." (Before you shake your head, can you think of a better name?) Take it as a matter of faith that the tau factor is independent of the air temperature used in the plunge test and is inversely proportional to the square root of the air velocity.

Since this mysterious tau factor is inversely proportional to the square root of the air velocity, it follows that if you multiply it by the square root of the air velocity at which it was measured, the result is a number which is independent of both the air temperature and air velocity. We end up with a number known as the Response Time Index (RTI). The units of this number are seldom mentioned since they are rather awkward. In the English system, they are sec ½ ft. ½. Since one foot = 0.3048 meters and the square root of 0.3048 is about 0.552, that factor can be applied to the RTI in the English system to get the SI equivalent. We will be using English RTI. Standard response sprinklers have RTIs ranging from 150 to 200 and sometimes they have been much higher, up to 400 or so. NFPA 13 defines fast-response sprinklers as having a thermal element of 50 s ½ meters ½ or less, which translates to about 90 s½ ft. ½, deviating from their practice of using English units for their definitions. It should be noted that a quick-response (QR) sprinkler is a specific type within the broad category of fast-response sprinklers, created by installing a fast-response link or bulb into what would otherwise be a standard spray sprinkler. NFPA 13 also defines standard response thermal elements as having an RTI of 80 s ½ m ½ (about 145 s ½ ft. ½) or more.

Having developed the fast-response sprinkler for residential use, attention then turned to incorporating the fast-response characteristic into sprinklers designed for high challenge fires. Traditionally, sprinkler systems were designed to control fires. In 1983, Factory Mutual launched what they called the Early Suppression Fast-Response (ESFR) sprinkler program. They put the "ES" ahead of the "FR" to emphasize the goal of suppression, rather than mere control. Two new terms were coined.

Required Delivered Density (RDD). The RDD is the minimum discharge density required to suppress a fire in a particular storage commodity at a given stage of the fire development. During the early critical stage of fire development, the rate of heat release increases rapidly, thus rapidly increasing the RDD.

Actual Delivered Density (ADD). The ADD is the amount of water actually delivered to the base of the fire at a given stage of fire development. When a sprinkler head is operating in the absence of a fire, all of the water being discharged, except for a very minor amount of evaporation, will reach the storage array. As a fire develops, the rate of evaporation increases, at some point, some of the droplets are converted to steam, and, most importantly, there is an ever-increasing updraft (which can reach 30 miles per hour, or more). The amount of water which can penetrate a given velocity of air depends upon the droplet size and the momentum (related to operating pressure plus gravity) of the drops. For the above two reasons, the ADD drops as the rate of heat release increases. We mention the droplet size. What determines that? It seems to depend upon the operating pressure and the size of the orifice. Research suggests that the median droplet diameter is roughly inversely proportional to the 113 power of the water pressure. For example, increasing the water pressure from 50 to 100 psi would decrease the mean droplet size by about 25%. It also appears that the mean droplet diameter is roughly directly proportional to the 213 power of the orifice diameter. Thus, comparing a 17/32 in. (k = 8) orifice to a 1/2 in. (k = 5.6) orifice, the larger orifice would increase the droplet diameter by about 5%. Comparing the original ESFR sprinkler (k = 1.4) to the standard 1/2 in. (k = 5.6) orifice, the larger orifice size increases the droplet diameter by about 18%. The significant characteristics are the relative surface area and the relative mass. Treating the droplets as spherical, an 18% increase in diameter means about a 39% increase in surface area and about a 64% increase in mass.

By definition, the Actual Delivered Density must exceed the Required Delivered Density at the time when sprinkler operation occurs if fire suppression is to be accomplished. Thus, the design density must be high enough so that the Actual Delivered Density stays above the Required Delivered Density during the time it takes for the sprinklers to respond. As we pointed out, the Actual Delivered Density is affected by the droplet size and the momentum of the drops. The ESFR sprinkler faithful to these principles was originally developed as a pendent sprinkler to maximize the momentum and the deflector was designed to maximize droplet size.

The rule for the shape of the design area for ESFR sprinklers is very simple. Assume 4 sprinklers to be flowing on each of 3 adjacent lines in the hydraulically remote area. The other rules for the ESFR sprinkler are also fairly simple.

The capabilities of the ESFR sprinkler are very impressive. Let's take a brief look at the advantages and disadvantages presented by ESFR sprinklers.

Advantages:
Minimizes fire damage through prompt suppression.
Tolerates a wide range of warehouse occupancy changes.
Eliminates the need for in-rack sprinklers.

Disadvantages:

Requires a very strong water supply. The design for a 30-foot high building requires over 1200 gpm with 50 psi pressure at the sprinklers. The design for a 40-foot building requires over 1450 gpm with 75 psi at the sprinklers. Many public water supplies will not provide these kinds of numbers. If the public water supply is inadequate, substantial expense is involved to provide the needed water supply and additional reliability problems are introduced.

Low tolerance for deviations from the rules. Typically, the performance of the first one or two sprinklers that open is critical to successful performance. Any deviation from the obstruction guidelines, for example, involving these initial sprinklers is likely to result in a catastrophic failure. The traditional "control" sprinkler systems are far more tolerant (maybe just a few more sprinklers will operate).

Traditional wisdom developed from experience with "control" sprinkler systems cannot be transferred to ESFR systems. It will be noted that the same rules apply to a Class 1 commodity as to plastic storage. If a system is designed for the far less demanding Class I commodity, surely the rules can be bent a bit. There must be a lot of overkill. Stop! It ain't necessarily so! The size of the fire when the first sprinkler operates is the critical consideration. Whatever the commodity, the fire has to reach a point where it is generating enough heat to fuse the sprinkler but not be too large for the sprinkler to be effective. This is a tricky business. Never think you are smart enough to tinker with the rules.

The 1996 Edition of NFPA 13 made a commendable effort to bring some degree of order out of what was, we think, correctly perceived as gradually evolving chaos.

We will take a brief look at some elements of the framework:

Types of sprinklers:

1. Upright and pendent spray sprinklers. These may be quick response except for extra hazard occupancies protected under the area/design method.
2. Sidewall spray sprinklers. These may be used only in light hazard occupancies with smooth, flat ceilings. An exception leaves the door open for ordinary hazard occupancies if a specific listing is achieved.
3. Extended coverage sprinklers. These are sprinklers listed for coverage exceeding normal rules. Construction must be smooth and unobstructed with the slope of the ceiling limited to not exceed 2 inches per foot. There is some further fine print here.
4. Open sprinklers. These are for deluge systems protecting special hazards, exposures, or other special locations, whatever they might be.
5. Residential sprinklers. These may be installed in dwelling units and adjoining corridors per the requirements of NFPA 130 or 13R.
6. Early Suppression Fast-Response sprinklers (ESFR). These are subject to construction limitations and wet systems unless a listing can be achieved to the contrary. Ceiling slope is limited to 2 inches per foot.
7. Large-drop sprinklers. There are special rules to minimize obstructions to the sprinkler discharge pattern. (Note: These sprinklers are now within a category known as Control Mode Specific Application or CMSA).
8. Quick-Response Early Suppression (QRES). This term was coined for a suppression sprinkler that could be developed for less than high challenge occupancies, although no such sprinklers have been developed to this point in time.
9. Special sprinklers. These are intended for the protection of specific hazards or construction features. To keep some semblance of order, performance criteria for listing are provided, orifice size and temperature ratings must fall into one of the established categories, and a restriction is placed upon the maximum area of coverage.

Occupancy-specific sprinklers. The development of sprinklers for narrow uses was prohibited. Exceptions were residential sprinklers and special sprinklers for protection of specific construction features. This was wise since, who knows, in the absence of this prohibition someone might decide to get a sprinkler listed for use in supermarkets. This would be a good source of revenue for the listing agencies but an unnecessary complication for most people who have to deal with sprinkler systems.

Orifice sizes. In general, a 50% flow increment when compared to the ½ in. (k = 5.6) diameter orifice sprinkler was required between diameter orifice sizes, increasing to 100% flow increments when the sprinkler k-factor exceeded 28 (1999 Edition of NFPA 13).

What the NFPA Committee on Automatic Sprinklers had intentionally advanced was the recognition that there were certain properties of sprinklers that could be arranged in many different ways to produce different protection options. The ESFR was a prime example, combining a fairly low temperature rating with low RTI to activate quickly in a fire, and with deflector characteristics and sufficiently large orifice to result in strong suppression capabilities.

Not all combinations, however, were considered necessary or practical. While a theoretical design approach was developed for the proposed Quick-Response Early Suppression (QRES) sprinkler[2] it was not pursued within NFPA 13 due to early recognition that it would not provide an economical alternative to the use of quick-response sprinklers using a traditional density/area approach. Although the high value of storage occupancies could justify the extra expense associated with an early suppression alternative, this was not true with ordinary non-storage occupancies. Fires were simply not frequent in most ordinary occupancies, and the performance of control-oriented sprinklers using the density/area approach was satisfactory. Once NFPA 13 began allowing the use of a reduced design area for quick-response sprinklers (strongly tied to ceiling height) in the 1996 Edition of the standard, it became clear that a QRES sprinkler design alternative would not be sought after.

The desire for economical yet effective fire protection has also created more options involving protection of storage occupancies in the past two decades.

Writing in SFPE's *Fire Protection Engineering* magazine in 2012,[3] Wes Baker of Factory Mutual, which by now had become FM Global, provided a summary of how that organization had begun to categorize sprinklers:

> By the start of the 21st century, sprinklers were commercially available in various K-factor sizes, orientations, nominal temperature ratings, RTI ratings, finishes and spacing coverage. They had been grouped into three categories, known today by the terms "control mode density area" (CMDA), "control mode specific application" (CMSA) and "suppression mode" (formerly called ESFR) sprinklers. The first two categories group sprinklers by an assumed performance during a fire event (i.e., control of a fire) whereas suppression mode sprinklers are assumed to suppress any fire that they protect. The assumed suppression performance allows for a reduced number of sprinklers in the design area (typically 12 sprinklers) as well as a reduced hose stream allowance (250 gpm [950 Lpm]) and sprinkler system duration (1 h). The CMDA sprinklers differ in design format as they utilize the density/area design format whereas both the CMSA and suppression mode sprinklers use the number of sprinklers at a given minimum pressure design format.

---

[2]Budnick, E. and Fleming, R., "Developing an Early Suppression Design Procedure for Quick Response Sprinklers," *Fire Journal*, National Fire Protection Association, November/December 1989.

[3]Baker, W., Jr., "The Whys Behind FM Global Data Sheets 2–0 and 8–9", *Fire Protection Engineering*, Society of Fire Protection Engineers, 2nd Quarter 2012.

While the 2019 Edition of NFPA 13 still specifically calls out design criteria for ESFR sprinklers, it also recognizes the terms CDMA and CMSA, and no longer contains the term "large-drop sprinkler". But this book is mainly about hydraulic calculations, and so it really doesn't matter what terms are used. The important thing is providing enough water for each sprinkler in the design area, whether it is a fixed minimum amount based on the listing of the sprinkler and its intended use, or a calculated amount based on the spacing and application of the sprinkler. Since the first density/area curves appeared in the 1972 Edition of NFPA 13, the Hazen–Williams formula has been in use for fire sprinkler calculations. And while a requirement to also apply the Darcy–Weisbach formula to large antifreeze systems was added in the 2007 Edition of the standard, the simpler Hazen–Williams formula has served the fire protection community extremely well for the past half-century.

We must close this section with an important caveat. We have tried to convey a general sense of what has been going on. Anyone who is responsible for designing or approving sprinkler systems must dig a lot deeper. The applicable codes must be thoroughly reviewed and understood. Much has been written and will be written by others which covers the subject matter in much more depth than fits the scope of this book. Further, the "state of the art" as we have tried to describe it may soon be out of date. Ongoing self-education is absolutely critical. Get involved and stay involved or stay away.

# What Are We Calculating?

Traditional hydraulically designed sprinkler systems involve a density and an area of sprinkler operation, also known as a design area or an area of application. The density/area design method is now part of the "Occupancy Hazard Fire Control Approach for Spray Sprinklers" within NFPA 13 and is also applicable to the use of Control Mode Density Area (CMDA) storage sprinklers. The most notable exceptions are CMSA and ESFR systems which we have just discussed, as well as Residential Sprinklers.

Density is measured in gallons per minute per square foot of floor area. The flow required from a sprinkler is determined by the area "covered" by the sprinkler multiplied by the desired density. NFPA 13 refers to the area "covered" as the "protection area." For example, if the sprinklers are spaced 10 feet apart along the branch lines and the branch lines are 12 feet apart, the protection area of each sprinkler is $10 \times 12 = 120$ ft$^2$. If the desired density is 0.20 gpm/ft$^2$, simply multiply 0.20 by 120. The product is 24 gpm, which means that all sprinklers in the design area must discharge at least this amount.

The determination of the minimum discharge per sprinkler sounds pretty simple but we need to look more closely. NFPA 13 provides the following guidance (introduced in the 1983 Edition, with minor changes since):

The protection area of coverage ($A_s$) per sprinkler shall be determined as follows:

1. Along branch lines: Determine the distance between sprinklers (or to wall or obstruction in case of the end sprinkler on the branch line) upstream and downstream. Choose the larger of either twice the distance to the wall or the distance to the next sprinkler. Define this dimension as "$S$".
2. Between branch lines: Determine the perpendicular distance to the sprinkler on the adjacent branch lines (or to a wall or obstruction in the case of the last branch line) on each side of the branch line on which the subject sprinkler is positioned. Choose the larger of either twice the distance to the wall or obstruction or the distance to the next sprinkler. Define this dimension as "$L$".
3. Protection area of the coverage of the sprinkler $= A_s = S \times L$.

Exception: In a small room (elsewhere defined as a room of light hazard occupancy classification, with unobstructed construction, not exceeding 800 ft$^2$),



H. S. Wass Jr. and R. P. Fleming P.E., *Sprinkler Hydraulics*,
https://doi.org/10.1007/978-3-030-02595-3_8

the protection area of each sprinkler in the small room shall be the area of the room divided by the number of sprinklers in the room.

These are sensible guidelines. While NFPA 13 should not feel called upon to say any more, we will return to this subject toward the end of this section to consider some of the configurations requiring judgment.

The protection area per sprinkler, measured in square feet, is the area over which the sprinkler is assumed to be discharging for purposes of the calculation. In reality, of course, the sprinkler will discharge over a larger area. The standard spray sprinkler, for example, is required to have a discharge not exceeding a circle 16 feet in diameter in a plane 4 feet below the sprinkler deflector when discharging 15 gpm.

Aside from the size of the design area, there is also the question of the shape and location of the design area. The required shape is a rectangle with certain parameters, and the location is the hydraulically most remote area. All of this is discussed later in the section on "The Hydraulically Most Remote Area."

Taking into account the discharge pattern of sprinklers, the "protection area" of a sprinkler is limited by some rules for spacing. For what NFPA 13 aptly defines as "light hazard occupancies" and "ordinary hazard occupancies", both the distance between sprinklers on branch lines and the distance between branch lines must not exceed 15 feet. For what they define as "extra hazard occupancies" and "high-piled" storage, the distance between sprinklers on branch lines is limited to 12 feet and the distance between branch lines is limited to 12 feet, or 12 feet 6 in. when there are 25 ft. bays. (Perhaps through oversight, this latter courtesy was not extended to extra hazard occupancies until the 1991 Edition.) The 1991 Edition also, for the first time, allowed 15 ft. spacing when the design density is less than 0.25 gpm/ft$^2$ without regard to occupancy class.

NFPA 13 also establishes "maximum sprinkler protection areas."

Light Occupancies: Hydraulically designed systems are allowed to take full advantage of the maximum IS-foot distance between sprinklers and branch lines, with a "protection area" limit of 225 ft$^2$ except for some types of combustible construction where lesser areas apply. Pipe schedule systems are limited to 200 ft$^2$.

Ordinary Hazard Occupancies: A 130 ft$^2$ limit applies in all cases.

Extra hazard occupancies and high-piled storage: The limit is 100 ft$^2$ except that it may be extended to 130 ft$^2$ when the design density is less than 0.25 gpm. Extra hazard pipe schedule systems are limited to 90 ft$^2$.

Having set forth the rules for sprinkler spacing and areas, we must now note that there is an exception to these rules. Listed extended coverage sprinklers may exceed the distance and area in accordance with the listings subject to a maximum protection area of 400 ft$^2$ for Light and Ordinary Hazard or 196 ft$^2$ for Extra Hazard and high-piled storage.

We have quoted the phrase "high-piled storage" without explanation. NFPA 13 defines it as "solid piled, palletized, rack storage, bin box, and shelf storage in excess of 12 ft. in height." This ties into the lower limits of what formerly fell under NFPA 231, *General Storage*, and NFPA 231C, *Rack Storage of Materials*, separate standards that were merged into NFPA 13 beginning with the 1999 Edition.

Of course, it was never really that simple. When the storage standards started addressing plastics, they rightly, in our view, went below the 12 ft. minimum storage height since plastics as little as 5 feet in height can have a rate of heat release which is "high challenge."

Where do we get the density and design area? NFPA 13 provides guidance for what they define as Light Hazard, Ordinary Hazard Groups 1 and 2, and Extra Hazard Groups 1 and 2 occupancies. (Prior to the 1991 Edition of NFPA 13, there were three Ordinary Hazard Groupings. In 1991, they combined Groups 2 and 3 into Group 2 and sort of split the difference on the density requirement. Not everyone agreed with this, but we think it was a sensible simplification.) As mentioned in the previous paragraph, NFPA 13 now also includes design criteria formerly found within separate NFPA standards on storage protection:

NFPA 231, *Standard for General Storage*
NFPA 231C, *Standard for Rack Storage of Materials*
NFPA 231D, *Standard for the Storage of Rubber Tires*
NFPA 231F, *Standard for the Storage of Roll Paper*

Some separate NFPA standards continue to contain their own sprinkler system design criteria, including

NFPA 30, *Flammable and Combustible Liquids Code*
NFPA 30B, *Code for the Manufacture and Storage of Aerosol Products*
NFPA 409, *Standard on Aircraft Hangars*

These three documents are unique in that the sprinkler system design criteria are so extensive that it could not easily be copied into NFPA 13 under the NFPA's extract policy. In the case of dozens of other NFPA codes and standards, however, such duplication was possible, and these criteria have been collected in a separate chapter within NFPA 13 entitled "Special Occupancy Requirements".

There are a few high challenge storage scenarios that are not addressed by NFPA standards. Carpet storage and hanging garment storage are two examples. FM Global and some other insurance organizations have published protection criteria for these.

One provision of the aircraft hangar standard merits discussion. Since its 1979 Edition, NFPA 409 has stated the following:

> Uniform sprinkler discharge shall be based on a maximum variation of 15% above the required discharge rate in gallons per minute per square foot.

Since every sprinkler must discharge the minimum flow, the total required flow is always more than the theoretical flow obtained by multiplying the density by the design area, and NFPA 409 limits the overage for each head to 15%. The question has arisen as to whether a similar constraint should apply to systems designed under other standards. NFPA 409 was mainly concerned with foam–water systems with a limited amount of the foam agent. For simple water systems, except where there is a limited water supply, uniform sprinkler discharge is probably not a real advantage if all sprinklers meet the minimum required discharge. The "overages" provide a small added safety factor in the event of a deterioration in the water supply. Of course, excess water beyond that required to control a given fire only contributes to the associated water damage, but since control should be the first priority, it is better to err on the side of too much water.

We promised to come back to the subject of the protection area, or area "covered" by each sprinkler. Consider the following:

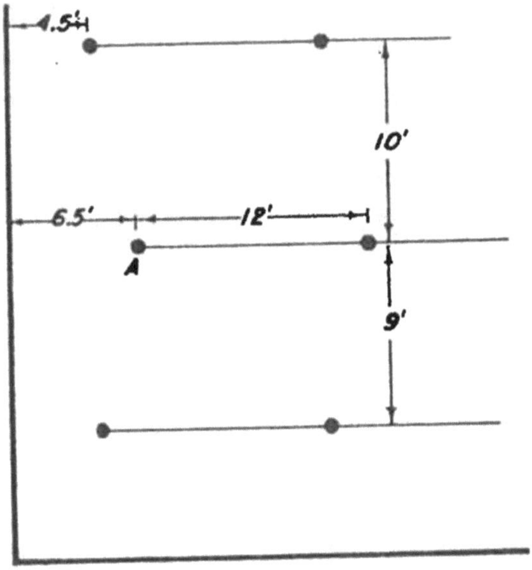


Area covered by Head $A$: Twice the distance to the wall is 13 ft. whereas the distance to the next head is 12 ft. so $S = 13$ ft. The larger distance to the next branch line is 10 ft. But the wording quoted earlier in this chapter from NFPA 13 says "Determine perpendicular distance to the sprinkler on branch lines." What does "perpendicular distance" mean? We know that perpendicular means "at right angles to." Right angles to what? It must be the branch line. If they meant the actual diagonal distance to the nearest sprinkler on the branch line (about 10.2 ft. in this example), they would not have said "perpendicular." So we will conclude $L = 10$.

$$S \times L = 13 \times 10 = 130\,\text{ft}^2$$

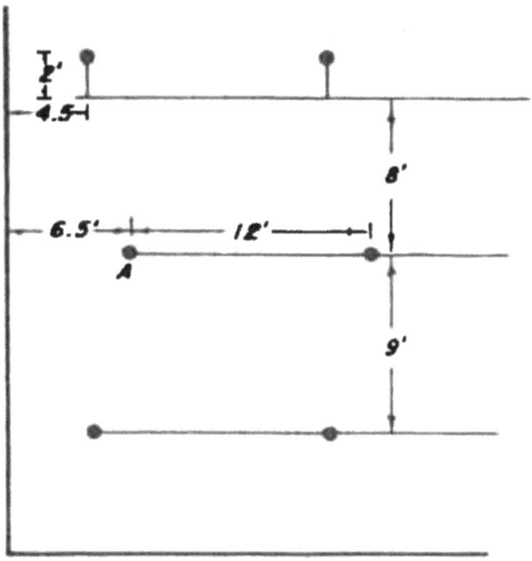


Prior to the 1989 Edition, NFPA 13 said "Determine the perpendicular distance to branch lines....on each side of branch line on which the subject sprinkler is positioned. Choose the larger..."

Strictly adhering to this wording, you would choose 9 ft. for the "*L*" in determining the protection area of Sprinkler *A*. If you look back to Page 27, you will note that the current wording is "Determine the perpendicular distance to the *sprinkler* on the adjacent branch lines...". This change, introduced in 1989, clarifies the intent. Obviously, it is the location of the sprinkler, not the branch line piping supplying it, that matters. Thus $L = 10$.

If it is agreed that it is the location of the sprinklers, not the location of the pipe that matters, it would seem to follow that the sprinkler-to-sprinkler distance on adjoining branch lines should be used, rather than the *perpendicular* distance. Before accepting this "logic", look at this.

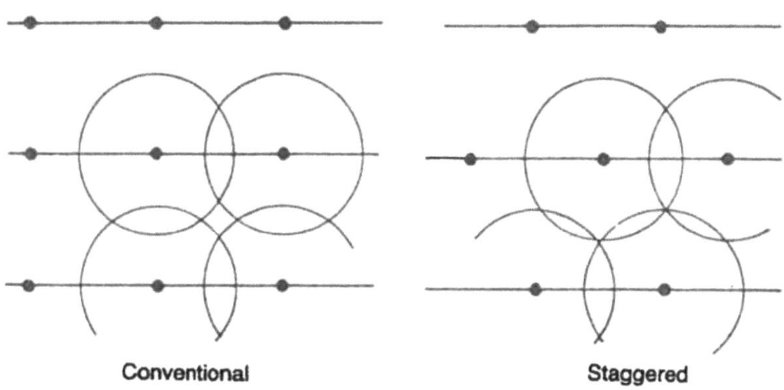

In terms of scale, these were drawn with a 10 ft. spacing between the sprinklers and a 10 ft. spacing between branch lines. The circles or arcs have a 6 ft. radius. It can be seen that the coverage is more uniform with the staggered sprinkler configuration and, in fact, NFPA 13 at one time required this configuration in certain instances. If sprinkler-to-sprinkler distance were used, the area "covered" by each sprinkler in the staggered arrangement would be 111.8 $ft^2$ versus 100 $ft^2$ in the conventional configuration. Despite the merit of this arrangement, overall there is one sprinkler per 100 $ft^2$ of floor area, and it would not seem reasonable to add an 11.8 $ft^2$ phantom area.

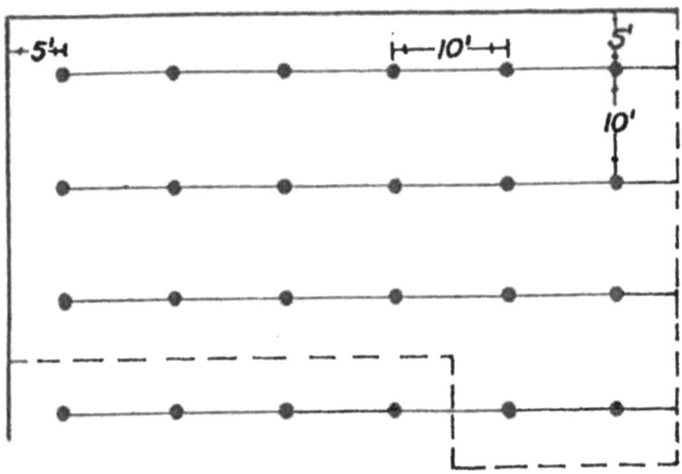


Let's look at something else:

We are using a design area of 2000 $ft^2$. The distance between sprinklers and between branch lines is 10 ft. Therefore, the design area for each sprinkler is $10 \times 10 = 100$ $ft^2$ and dividing 2000 by 100, we determine that there must be 20 sprinklers in the design area. Now suppose we move the wall at the top out one foot.

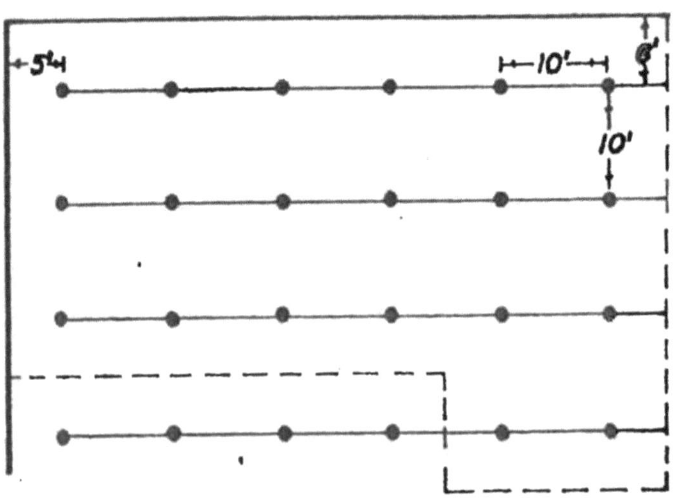

Now the distance from the last branch line to the wall is 6 feet rather than 5 feet. To quote again the applicable part of the rule, "Choose the larger of (1) the larger distance to the next branch line, or (2) in the case of the last branch line, twice the distance to the wall. Call this '*L*'." Twice the distance to the wall now becomes 12 and the design area for each of the six sprinklers on the end branch line becomes $10 \times 12 = 120$ ft$^2$ $6 \times 120 = 720$ which leaves 1280 ft$^2$ to be handled by the sprinklers on the other branch lines (13 sprinklers), or does it? The actual physical area is $10 \times 11 \times 6 = 660$ ft$^2$. Although some argued that the "phantom area" of 60 ft$^2$ (720–660) should count toward meeting the required design area, commentary first appearing in the 1996 Edition of the NFPA Handbook specifically refuted the notion. Therefore, the remaining 1340 ft$^2$ requires 14 sprinklers since the protection area of these sprinklers is 100 ft$^2$ and the design area remains the same as in the previous example.

This issue was settled with the introduction of a new requirement in the 2013 Edition of NFPA 13:

"23.4.4.1.1.5 Where the total design discharge from these operating sprinklers is less than the minimum required discharge determined by multiplying the required design discharge density times the required minimum design area, an additional flow shall be added at the point of connection of the branch line to the cross main furthest from the source to increase the overall demand, not including hose stream allowance, to the minimum required discharge as determined above."

A newcomer to this subject will wonder why we have six sprinklers per line in a design area. We will get to that in the section on "The Hydraulically Most Remote Area."

Let us move the wall the other way, with only a 4-foot distance to the end branch line.

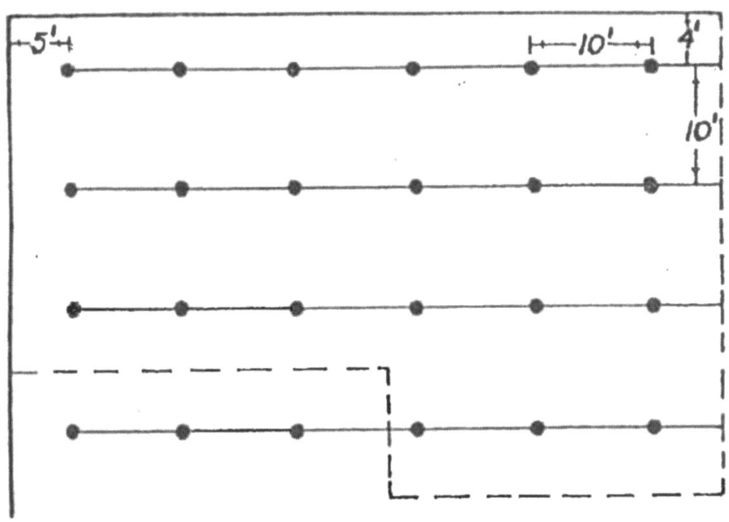


The "*L*" for the last branch lines remains 10 as it was in the initial example, since the distance to the next branch line is larger than twice the distance to the wall. Thus, the design area for the sprinklers on the last branch line remains 100 ft$^2$ and we still seem to need 20 sprinklers. But the actual building design area is $29 \times 60 = 1740$ plus $2 \times 100 = 200$, or 1940 ft$^2$.

What are we to make of this? We are not sure but a case can be made for blindly following the "$S \times L$" rule and not worrying about it. One highly regarded sprinkler guru, Russ Fleming of the National Fire Sprinkler Association, took a look at this in the Winter 1990 Edition of that association's *Sprinkler Quarterly.* He pointed out that the NFPA's *Automatic Sprinkler System Handbook* took the position we suggested above, then took the opposite view in the 1989 Edition. Russ then went on to argue that walls "tend to assist the sprinkler system by limiting the number of sprinklers that can operate" and goes into some detail as to why this is probably so. Therefore, he concluded that "it should not be of great concern if, due to some sprinklers being close to the end walls, the floor area covered by sprinklers in the hydraulically most remote area is somewhat less than the area initially selected from the area/design curve."

However, the 1996 *Sprinkler Systems Handbook,* in its explanatory material, stated that "this deficit must be accounted for", which in the example above would mean that the 60 ft$^2$ "deficit" calls for a 21st sprinkler in the design area. The NFPA 13 Committee resolved this in the 2013 Edition when dealing with the issue of small design areas for higher hazards, adding a section to require that "Where the total design discharge from these operating sprinklers is less than the minimum required discharge determined by multiplying the required design density times the required minimum design area, an additional flow shall be added at the point of connection of the branch line to the cross main furthest from the source to increase the overall demand, not including hose stream allowance, to the minimum rerquired discharge as determined above."

We will now take a look at another ambiguous issue that arises when you have sprinklers beneath a pitched roof. Starting on Page 97 we have an example of calculations for such a case. This appeared in the first edition of this book in 1983. We have never been challenged on this. Ed Miller of the American Fire Sprinkler Association discussed this subject in the October 1993 Edition of their publication *Sprinkler Age.* He had been consulted by a sprinkler contractor who had a disagreement with an Authority Having Jurisdiction.

Until the 1991 Edition, NFPA 13 said that "density shall be calculated on the basis of floor area." In 1991, the wording was changed to "the density shall be calculated on the basis of area of sprinkler operation. The area covered by any sprinkler…shall be determined in accordance with 4-2.2.1." 4-2.2.1 is the "protection area" "*S*" and "*L*" rule cited at the beginning of this section.

NFPA 13 says that "the distance between sprinklers…shall be measured along the slope" and has said that for many years with slightly different wording. Obviously, this "distance" relates to the maximum and minimum allowable distance between sprinklers. It also seems reasonable to suppose that this "distance" should be plugged into the *S* and *L* rule for determining the "protection area per sprinkler" which, in turn, must comply with the maximum allowable protection area previously discussed. As Ed Miller puts it, "the question arises of whether to use area of coverage per sprinkler on the slope or the area of coverage projected on the floor in determining the remote area and density to be used for the hydraulic calculations." Not unreasonably, considering the wording in NFPA 13, he tilted toward using the area of coverage on the slope, which happens to be more demanding because the larger area requires a larger flow from each sprinkler.

Possibly in response to this, the word "floor" was added back into 6-4.4.3 in the 1994 Edition of NFPA 13:

> The density shall be calculated on the basis of floor area of sprinkler operation. The area covered by any sprinkler used in hydraulic design and calculations shall be the horizontal distances measured between the sprinklers on the branch line and between the ·branch lines in accordance with paragraph 4-2.2.1.

This still wasn't clear enough, however, so the issue of calculations for sloped ceilings and roofs was further clarified in the 2007 Edition of NFPA with the addition of Section 22.4.4.5.6 and accompanying annex material. The new section read as follows:

> For sloped ceiling applications, the area of sprinkler application for density calculations shall be based upon the projected horizontal area.

Annex guidance explained that for the common situation in which the slope runs parallel to the branch lines, a calculation could be made to determine the projected area per sprinkler for hydraulic calculation purposes:

$$A_s = S' \times L$$

where

$S'$ $(\cos\theta)\,S$
$\Theta$ the angle of the slope
$S$ the distance between sprinklers on the branch line

Before leaving the subject of sprinkler spacing, look at this real-life schematic, which can be found in the headquarters of the National Fire Protection Association (unless the area has been remodeled).

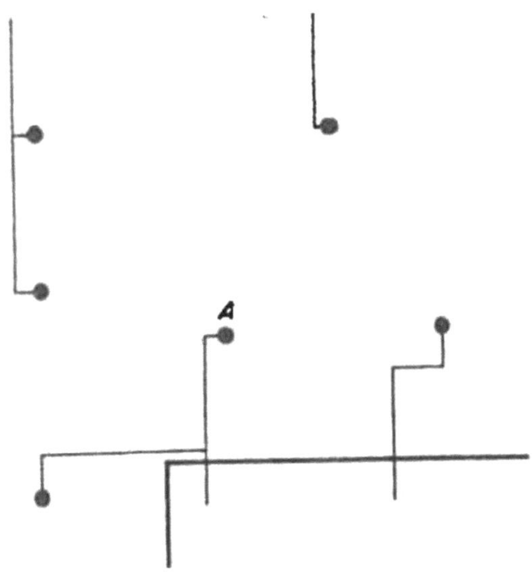

How do you determine the area covered by Sprinkler *A*? We are open to suggestions, but here is our approach. Look at this without the piping.

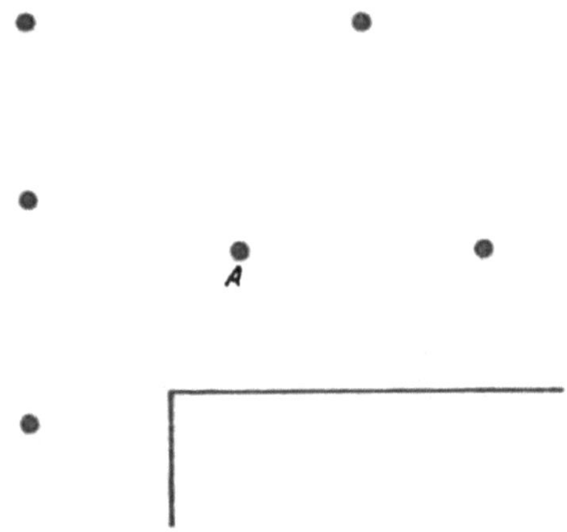


Visualize parallel branch lines. Suppose the branch lines were arranged like this.

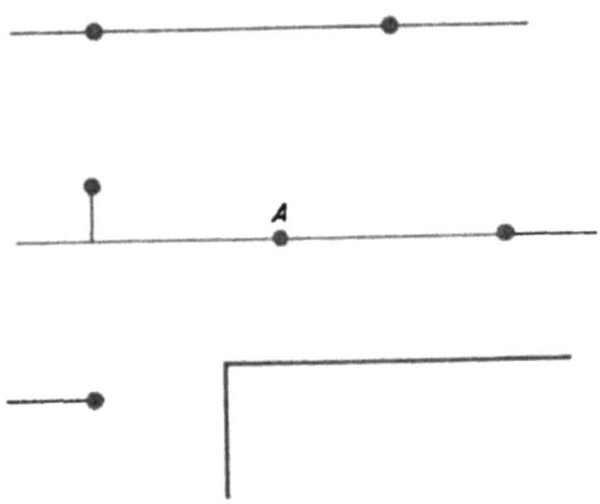


In this configuration, it is a simple matter to apply the rules that have been set forth.

Up to now, we have been discussing the so-called "Area/Density Method." While not commonly used, NFPA 13 has long sanctioned what they call the "Room Design Method." While it makes a certain amount of sense, its use is not universally sanctioned and we have some reservations.

The wording of the room design method is somewhat awkward so we will attempt to describe it a little differently. Instead of the prescribed rectangle with an area set forth in the area/density curves, the design area is the room that creates the greatest demand. For light and ordinary hazard occupancies, the minimum area in the area/density curves is 1500 ft$^2$. Most buildings have a "room" larger than this, in which case, while the room design method may still be an option, it is academic.

Where you do have a high degree of compartmentation throughout and you want to consider the room design method because your total sprinkler water demand would be less, it is necessary to look at what qualifies as a "room."

According to NFPA 13, "To utilize the room design method, all rooms shall be enclosed with walls having a fire resistance rating equal to the water supply duration indicated in Table 19.3.3.1.2 (2019 Edition)". The Annex states that "walls may terminate at a substantial suspended ceiling", leaving it up to the Authority Having Jurisdiction to decide what "substantial" means. The referenced water supply durations are as follows:

| | |
|---|---|
| Light hazard | 30 min |
| Ordinary hazard | 60–90 min |
| Extra hazard | 90–120 min |

Reading further, you will find that the lower duration values provided for Ordinary and Extra Hazard apply only where sprinkler system waterflow alarm device(s) and supervisory device(s) are electrically supervised and such supervision is monitored at an approved, constantly attended location. In the absence of such supervision, the higher values should be used. That covers the walls. What about the doors or other openings? For Ordinary and Extra Hazard occupancies, automatic or self-closing doors "with appropriate fire resistance rating for the enclosure" are required. While the provision is there, it would be an unusual Ordinary or Extra Hazard Occupancy that qualified for the room design method. This method is really intended for the Light Hazard occupancy where a high degree of compartmentation is more common. In a Light Hazard occupancy, no doors are required. In the absence of "automatic or self-closing doors" (no fire resistance rating required), two sprinklers in the communicating space nearest each unprotected opening (or one sprinkler if the communicating space has only one sprinkler) must be included in the calculations. "The selection of the room and communicating space sprinklers to be calculated shall be that which produces the greatest hydraulic demand."

While requirements are set forth for walls and horizontal openings, there is no mention of vertical protection if you have a multi-story building.

Presumably, it is felt that the sprinklers preclude the need for any kind of vertical fire resistance rating. The absence of any mention of vertical protection does not mean, however, that common sense should be held in abeyance. If there were some kind of unprotected vertical opening in the room, perhaps an open dumb waiter, we suggest that all bets are off. Do not use the room design method.

There is one other bit of guidance on the room design method. If "the area under consideration is a corridor protected by a single row of sprinklers with protected openings in accordance with,... the maximum number of sprinklers that needs to be calculated is five or, when extended coverage sprinklers are installed, all sprinklers contained within 75 linear feet of the corridor."

Where the openings for the corridor are not protected, a five sprinkler design area is nevertheless permitted in Light Hazard occupancies, with the same 75 linear foot rule applicable to the use of extended coverage sprinklers.

In other occupancies, the Room Design Method cannot be applied with unprotected openings from a corridor, but a "special design area" rule applies to any single line of sprinklers, requiring the design area to include all sprinklers on the line to a maximum of seven.

The applicable density when using the room design method comes from the appropriate area/density curve, using the minimum area when the "room" area is less than the minimum area.

While the room design method has some logic in a highly compartmented building, it introduces another element of uncertainty over the life of the building. It is not unusual to make changes in these kinds of buildings. Walls may be moved or eliminated. When walls are changed in buildings where the area/density method has been used, the main concern is with the distance between the existing sprinklers and the walls. In most cases, this can be easily noted and evaluated by anyone with basic sprinkler knowledge. With the room design method, a careful review of the original sprinkler design is needed. This information may be unavailable and it may be falsely assumed that the area/density method was used. As with too many other elements of sophisticated sprinkler design, the theoretical underpinnings are perfectly sound but there is no practical mechanism for assuring that the design scenario will still be in place when the fire occurs five years later.

Before we leave this section, we will also mention what is called a "special design approach" pertaining to residential sprinklers. While, as discussed earlier, residential sprinklers were developed primarily to provide economical life safety systems for residential occupancies and NFPA 13D and 13R set forth the rules for those installations, NFPA 13 recognized that residential sprinklers were preferable in residential portions of building being protected in accordance with NFPA 13. They inferentially make clear that residential sprinklers, and the residential "design approach" only applies to the residential areas by saying that "where areas such as attics, basements, or other types of occupancies are outside of dwelling units but within the same structure, these areas shall be protected" under the normal provisions of NFPA 13. The annex clarifies that corridors associated with apartment units are treated as being within the "residential area."

Within the residential area, the design area shall be "that area that includes the 4 hydraulically most demanding sprinklers." It is also specified, of course, that sprinkler discharge rates meet the individual sprinkler listing requirements.

Let us consider a typical residential building with apartments on three floors and a combustible attic. If sprinklers are provided, typically it would be designed in accordance with NFPA 13R with no sprinklers in the attic. If, however, there was a concern for property protection as well as life safety, an NFPA 13 system might be installed. The attic sprinklers would be designed in accordance with area/density method, or possibly the room design method if the attic space were broken up as such attic spaces should be. A separate set of calculations would be required for the residential area, involving the 4 most hydraulically demanding residential sprinklers.

Some specific building areas have special design areas. One is a minimum design area of three sprinklers that applies to a building service chute supplied by a separate riser, with each sprinkler required to have a minimum discharge of 15 gpm. Another is that sprinklers located within ducts are required to flow at a minimum discharge pressure of 7 psi, with all sprinklers in the duct flowing simultaneously.

# Discharge from a Sprinkler

It is time to look at what might be called the "bottom line" of this book—what comes out of the sprinkler. As common sense suggests, what comes out is a function of the size of the opening (more pretentiously known as the orifice), of the physical characteristics in the vicinity of the opening, and of the pressure.

The theoretical flow through an orifice can be expressed in terms of velocity and cross-sectional area:

$$Q = av$$

where $Q$ is the flow in cubic feet per second
$a$ is the cross-sectional area in square feet
$v$ is the velocity in feet per second.
Since

$$a = \frac{\pi D^2}{4}$$

or, converting the diameter from feet to inches,

$$a = \frac{\pi \left(\frac{D}{12}\right)^2}{4} = \frac{\pi D^2}{576}$$

where "*d*" the diameter, is in inches.

Reference to the section on velocity pressure (Page 73) reveals that

$$velocity\,head = \frac{v^2}{2g}\,and\,pressure\,head = \frac{p}{w}$$

where $p$ is pressure, in pounds per square foot and $w$ is weight, in pounds, of a cubic foot of freshwater.



H. S. Wass Jr. and R. P. Fleming P.E., *Sprinkler Hydraulics*,
https://doi.org/10.1007/978-3-030-02595-3_9

Converting to pounds per square foot,

$$\frac{p}{\frac{62.4}{144}} = \frac{144}{62.4}p$$

where $p$ is now pressure in psi.

Theoretically, when water is discharged through an orifice, the pressure head is converted into a velocity head.

Therefore,

$$\frac{144}{62.4}p = \frac{v^2}{2g}, \text{thus } v^2 = \frac{144}{62.4} \times 2 \times 32.2p$$

and

$$v = \sqrt{\frac{144}{62.4} \times 2 \times 32.2p}$$

Thus

$$Q = av = \frac{\pi d^2}{576}\sqrt{\frac{144}{62.4} \times 2 \times 32.2p}$$

where $Q$ is in cubic feet per second, but we want $Q$ in terms of gallons per minute. There are about 7.4805 gallons in a cubic foot and 60 seconds in a minute, so

$$Q = 60 \times 7.4805 \times \frac{\pi d^2}{576}\sqrt{\frac{144}{62.4} \times 2 \times 32.2p} = 29.84d^2\sqrt{p}$$

where $Q$ = gpm, $d$ = diameter, in inches, and $p$ = pressure, in psi.

Since actual, as opposed to theoretical, discharge from an orifice is affected by friction, turbulence, and contraction of the stream, a discharge coefficient, "$c$," must be added, yielding

$$Q = 29.84cd^2\sqrt{p} \tag{1}$$

29.84 $cd^2$ can be reduced to a single constant, "$k$," for a given sprinkler and the formula for calculating the discharge from a sprinkler becomes

$$Q = k\sqrt{p} \tag{2}$$

where

$Q$ discharge in gallons per minute (gpm)
$k$ the discharge coefficient, a constant
$p$ water pressure in pounds per square inch (psi).

Strictly speaking, the "*k*" varies slightly with the pressure, but this can normally be ignored. As a matter of interest, the published "*k*" for a sprinkler is probably about right in the 40–45 psi range. The actual "*k*" is slightly higher at low pressures and slightly lower at high pressures. To cite an actual example, a sprinkler that has a published "*k*" *of 5.53* ranges from 5.7 at 5 psi to 5.4 at 100 psi. An increase in the "*k*" yields a higher flow, but a lower pressure. Conversely, decreasing the "*k*" results in a lower flow and a higher pressure. For example, the design area illustrated on page 108 requires a flow of 708.9 gpm at *85.6* psi at the point shown to deliver the specified 0.30/2000, using a "*k*" of 5.6. The following table shows the effect of a small change in the "*k*":

| *k* | *Q* | *p* |
|---|---|---|
| 5.5 | 705.2 | 86.3 |
| 5.6 | 708.9 | 85.6 |
| 5.7 | 712.0 | 84.9 |

If these flows were carried back to the base of the riser, there would be a very slight convergence of the pressures since the higher flows would generate higher friction losses. Related to most water supplies, the pressure differences shown above are more significant than the flow differences; thus the calculations for a low-pressure system would tend to be conservative (and the reverse would be true of a high-pressure system) since the actual "*k*" at the operating pressure would probably be higher than the "*k*" at which it was calculated.

The discharge coefficient, "*k*," varies slightly for different sprinklers. However, NFPA 13 requires the use of nominal *k*—factors in calculations:

5.6 for the standard 1/2-in. orifice sprinkler,

8.0 for the 17/32 in. "large" orifice sprinkler,

11.2 for the 5/8 in. "extra-large" orifice sprinkler, and

14.0 for the 3/4 in. "very extra-large" orifice sprinkler, with newer larger orifice sizes standardized at k-factors of 16.8, 19.6, 22.4, 25.2, and 28.0, each step representing an increase in flow corresponding to the flow of a 5.6 sprinkler.

The difference in the "*k*" values for the two historically most common sprinklers, standard and large orifice, may seem puzzling. The discharge formula tells us that the "k"s should vary according to the ratio of the squares of the orifice diameters, yet the ratio of the squares of 1/2 and 17/32 is about 1.13 whereas the ratio of the "*k*"s is about 1.43. Some nominal 1/2-in. sprinklers have a tapered nozzle with a diameter of about 7/16 in. at the discharge point, but, on the other hand, some have a full 1/2 in. discharge orifice. In any case, listed sprinklers must be designed so that the "*k*" falls between 5.3 and 5.8 for the 1/2 in. sprinkler and 7.4 and 8.2 for the 17/32 in. sprinkler. Further, the production tolerance must be such that the "*k*" does not vary more than ±5%.

There is more to be said about the "*p*" in the discharge formula, but that will wait until the section on velocity pressure.

Before leaving this subject, it should be noted that the pressure at the sprinkler is of concern for more reasons than the resultant flow. The pressure affects the spray pattern and the droplet size. Droplet size is important because it relates to the ability of the sprinkler discharge to penetrate the fire plume and reach the seat of the fire. Fire tests suggest that for storage occupancies, the most favorable droplet size may occur in the range from 30 to 60 psi. Unless there is a low ceiling, there is a reason for concern about the ability of the fine droplets to penetrate the updraft in a fire when the discharge pressure is above 100 psi. The 1999 Edition of NFPA 13 for the first time required, for standard response spray sprinklers, a minimum nominal *k*-factor of 8 for storage unless the density is 0.20 gpm/sq. ft. or less and a minimum nominal *k*-factor of 11 when the density exceeds 0.34 gpm/sq. ft. The stated reason is to limit pressures to avoid an unfavorable "misting effect." The only reference to a minimum pressure appears in NFPA 13. The minimum allowable pressure, which applies to *all* sprinkler systems, is 7 psi. Prior to the 1996 Edition of NFPA 13, there were some sprinklers listed for minimum operating pressures below 7 psi. Concerns about the distribution pattern and concerns about the ability of a lower pressure to blow off the orifice cap some years after installation led to a 7 psi minimum operating pressure requirement in the 1966 Edition of NFPA 13. While some sprinklers are listed with higher minimum operating pressures, there are no longer sprinklers listed with a minimum pressure below 7 psi.

# Elevation Changes

Changes in elevation must *always* be taken into account in calculating a sprinkler system.

All of us(?) have learned somewhere along the way that a cubic foot of water weighs 62.4 lb. (Actually, 62.4 is the weight of water at 52.72°F. It weighs slightly less at higher temperatures, slightly more at lower temperatures. 52.72°, however, is a reasonable temperature).

If you think of a cubic foot as a cube one foot square and one foot high, it should be apparent that when you divide 62.4 by the number of square inches in a square foot, the result is the weight of a one-square-inch column of water one foot high, or the pressure, in pounds per square inch, exerted by one foot of water.

$$62.4 \div 144 = 0.433$$

Thus, the pressure must be increased by 0.433 psi for every foot by which the elevation is reduced and decreased by 0.433 psi for every foot by which the elevation is increased. When all sprinklers assumed to be discharging are at the same elevation, it is recommended that all elevation changes be ignored until you reach the main riser. At the main riser, the pressure should be increased to reflect the total difference in elevation between the base of the riser (see definition on page 102) and the sprinklers assumed to be operating.

When the sprinklers assumed to be discharging are at different elevations (a pitched roof is a common example—see pages 97 through 100), it is necessary to adjust the pressure for elevation at each step in the assumed discharge area because the pressure change due to elevation will affect the actual discharge. Again, however, all elevation changes not affecting a discharging sprinkler are best ignored until the calculations reach the main riser. It is simpler to account for all elevation changes at one time and, when branch-line $k$'s are used, unnecessary errors are avoided (see page 185).



H. S. Wass Jr. and R. P. Fleming P.E., *Sprinkler Hydraulics*,
https://doi.org/10.1007/978-3-030-02595-3_10

# Sprinkler Piping—Nothing Is Simple These Days

Prior to about 1970, Schedule 40 steel pipe was used in virtually all sprinkler systems. Life was simple.

In the early 1970s, Schedule 10 pipe, commonly referred to as thin-wall pipe, was introduced. The original edition of this book, published in 1983, contained pertinent data such as internal diameters for Schedule 40 and Schedule 10 pipe plus copper tubing. Copper tubing also appeared in the 1970s although its use remains very limited to this day. That pretty well covered the subject at that time.

In the early 1980s, Allied Tube and Conduit introduced a threadable lightwall pipe known as XL. In the subsequent years, a wide range of pipe products emerged, including other lightwall and ultralight wall pipes. Associated with the piping, of course, are a variety of joining methods and fittings. Joining methods, in addition to the traditional threaded connections associated with Schedule 40 pipe, include cut grooved, rolled grooved, plain end, special listed fittings, and welding. There is a need for awareness of what kind of joining methods are acceptable for the particular kind of pipe being used.

Reportedly, pipe constitutes about one-third of the total cost of a sprinkler installation so there is a strong incentive to seek the lowest cost pipe.

NFPA 13 addresses the subject in terms that have little meaning to the casual reader.

> Pipe or tube shall meet or exceed one of the standards in Table 7.3.1.1 or be in accordance with 7.3.3.

Table 7.3.1.1 lists some ASTM and ANSI standards. Section 7.3.3 includes this statement:

> Other types of pipe or tube investigated for suitability in automatic sprinkler installations and listed for this service, including steel, and differing from that provided in Table 7.3.1.1 shall be permitted where installed in accordance with their listing limitations, including installation instructions.



H. S. Wass Jr. and R. P. Fleming P.E., *Sprinkler Hydraulics*,
https://doi.org/10.1007/978-3-030-02595-3_11

While nonmetallic CPVC pipe originally fell into the category of "other", it is now included in Table 7.3.1.1. Prior to such inclusion, an appendix/annex section contained some cautionary language regarding nonmetallic piping:

> With respect to thermoplastic pipe and fittings, exposure of such piping to elevated temperatures in excess of that for which it has been listed may result in distortion or failure. Accordingly, care must be exercised when locating such systems to ensure that the ambient temperature, including seasonal variations, does not exceed the rated value.

The annex section A.7.3.3 now includes lightweight steel pipe as an example of "other types", and continues to make some vague cautionary comments concerning special listed piping products:

> While these products can offer advantages, such as ease of handling and installation, cost effectiveness, and reduction of friction losses, it is important to recognize that they also have limitations that are to be considered by those contemplating their use or acceptance.
>
> Corrosion studies have shown that, in comparison to Schedule 40 pipe, the effective life of lightweight steel pipe can be reduced, the level of reduction being related to its wall thickness. Further information with respect to corrosion resistance is contained in the individual listings for such pipe

Some years back when approval of threaded thin-wall pipe was sought, the issue of corrosion resistance came to the forefront because of the reduced thickness of the threaded section of the pipe. Underwriters Laboratories developed a "Corrosion Resistance Ratio" (CRR) which is a factor indicating the expected life of a steel piping product relative to threaded Schedule 40 pipe.

Underwriters Laboratories defines the CRR as follows:

$$CRR = \frac{(X)^3}{X_{40}}$$

$X_{40}$ is the thickness of Schedule 40 pipe under the first exposed thread. The "first exposed thread" is the minimum pipe thickness exposed to both interior and exterior corrosion and occurs at the threaded joint assembly at a line defined by the thread width, just before the pipe engages the fitting.

$X$ is the thickness of the Listed pipe measured either under the first exposed thread for threaded pipe or at the thinnest wall section for unthreaded pipe.

The CRR does not, however, give you any idea of the life expectancy of a steel piping product in a particular installation. The corrosivity of the water supply and the atmosphere in which the pipe is being used are the critical variables. Also, the CRR is based upon the relative wall thickness and does not take into account possible differences in the corrosion rate in the newer kinds of pipe, which have different alloys to obtain higher tensile strengths. The cautionary comments in NFPA 13, quoted on the last page, are simply suggesting that all piping listed by Underwriters Laboratories may not be suitable for relatively corrosive environments. Judgment should be exercised since an installation that fully complies with NFPA 13 may not always be appropriate. One knowledgeable and reputable person has suggested in print that while at one time it was felt pipe should be adequate for

50 years' service, "technological obsolescence" of buildings may now mean that a life of 20–30 years is adequate. We disagree. In fact, there are many sprinkler systems in service today that are more than 50 years old and functioning just fine. To install a pipe that might start falling apart in 20 or 30 years is, we think, not responsible. Of course, the "guilty" parties may not be around at that time so the temptation to come in with the lowest price may be hard to resist.

As indicated above, the most common type of plastic pipe, approved for use in sprinkler systems in residential and Light Hazard occupancies, subject to numerous requirements and restrictions, is post-chlorinated polyvinyl chloride (commonly referred to as CPVC). Plastic pipe has some hydraulic and corrosion advantages, but also has some drawbacks as compared to steel pipe:

1. It is combustible. Consequently, it is only permitted in wet pipe systems and, in most cases, must be protected against fire exposure.
2. It cannot withstand much heat. CPVC pipe must be limited to an exposure of 150 °F.
3. It has less mechanical strength.
4. It has a much higher thermal expansion coefficient. For example, consider a 50-foot straight length of pipe and a 25 °F temperature change. The change in length would be about 0.10 in. for a steel pipe but 0.51 in. for CPVC pipe. Thermal expansion and contraction can be an installation consideration where significant temperature variations may take place.

Copper tubing is also approved for sprinkler systems, subject to appropriate specific installation requirements. Types K, L, and M copper tubing are acceptable but Type M is normally used because it has the thinnest wall, the largest inside diameter and is the least expensive.

# Friction Loss of Water Flowing in a Pipe

Antoine de Chezy (1718–1798), a French engineer who conducted studies in connection with the construction of canals, pioneered the development of a formula for computing friction loss of water flowing in an open channel or in a pipe:

$$v = c\sqrt{rs}$$

where

$v$ average velocity of flow,
$c$ coefficient reflecting roughness of pipe,
$r$ internal hydraulic radius, equal to

$$\frac{area}{circumference} = \frac{\frac{\pi d^2}{4}}{\pi d} = \frac{d}{4}$$

$s$ hydraulic slope, which may be considered friction loss per unit length.

Another French engineer, Henri-Philibert-Gaspard Darcy (1803–1858), while designing and constructing a municipal water-supply system in Dijon, ran tests on flow in pipes. As a result of his work and the work of Julius Weisbach, J. T. Fanning, and others, the friction factor in de Chezy's equation was refined and expressed as the following equation, commonly known as the Darcy–Weisbach equation:

$$h = f \times \frac{1}{d} \times \frac{v^2}{2g}$$

where

$h$ head loss due to friction
$f$ friction factor



H. S. Wass Jr. and R. P. Fleming P.E., *Sprinkler Hydraulics*,
https://doi.org/10.1007/978-3-030-02595-3_12

*l* length of pipe
*d* internal diameter of the pipe
*v* flow velocity
*g* acceleration due to gravity.

The friction factor, "*f*", is difficult to calculate, and it remained difficult for Gardner S. Williams and Allen Hazen to produce, in 1905, a version of the de Chezy formula that has been adopted for general use in fire protection.

$$v = Cr^{0.63}s^{0.54}0.001^{-.04}$$

The units must be transposed to put this equation into a form in which it can be conveniently used.

*c* is the Hazen–Williams coefficient
*r* is the hydraulic radius (see preceding page), which is equal to *d*/4 where d is in feet or *d*/(4 × 12) = *d*/48 where *d* is in inches
*s* is the hydraulic slope, or the friction pressure drop, in feet, divided by the pipe length. The pipe length will be considered to be one foot to yield an equation for friction loss per foot. Referring to the section on elevation changes (Page 45),

$$psi = \frac{62.4}{144} \times height, in feet\ or\ h = \frac{144}{62.4}p\ where\ p\ is\ in\ psi.$$

*v* is velocity, in feet per second.

Referring to the section on velocity pressure,

$$v = \frac{4 \times 144Q}{60 \times 7.405\pi d^2} = \frac{576Q}{448.83\pi d^2}$$

Substituting, the basic Hazen–Williams equation becomes

$$\frac{576Q}{448.83\pi d^2} = C \times \left(\frac{d}{48}\right)^{0.63} \times \left(\frac{144}{62.4}p\right)^{0.54} \times 0.001^{-0.04}$$

Solving for *p* (friction loss, in psi, per foot),

$$\left(\frac{144}{62.4}p\right)^{0.54} = \frac{\frac{576Q}{448.83\pi d^2}}{C \times \left(\frac{d}{48}\right)^{0.63} \times 0.001^{-0.04}} = \frac{576Q \times 0.001^{0.04}}{448.83\pi d^2 \times C \times \left(\frac{d}{48}\right)^{0.63}}$$

$$\frac{144}{62.4}p = \frac{(576)^{1/0.54} \times Q^{1/0.54} \times 0.001^{0.04/0.54}}{(448.83\pi)^{1/0.54} \times d^{2/0.54} \times c^{1/0.54} \times \left(\frac{d}{48}\right)^{0.63/0.54}}$$

$$p = \frac{62.4}{144} \times \frac{129387.8474 \times Q^{1.85} \times 0.5994842503}{679063.9085 d^{3.70} \times C^{1.85} \times \frac{d^{1.17}}{91.50569213}}$$
$$= \frac{4.529297259}{C^{1.85} d^{4.87}} \times Q^{1.85}$$

Note that we have rounded off the exponents to two decimal places but have carried the other numbers considerably further.

NFPA 13, providing guidance for calculating friction loss, states the Hazen–Williams equation as follows:

$$p = \frac{4.52}{C^{1.85} d^{4.87}} \times Q^{1.85} \tag{3}$$

where

$p$ = friction loss per foot of pipe in pounds per square inch
$d$ = internal pipe diameter in inches
$Q$ = flow in gallons per minute
$C$ = Hazen–Williams coefficient.

Other numbers, such as 4.54, 4.524, and 4.576, have appeared authoritatively in the past. The difference, presumably, mainly results from rounding off the various numbers. The NFPA 13 number, 4.52, is the one generally used and we recommend its use, even though one of the other numbers might be equally valid.

Just to keep things a bit confusing, note that the 1.85 power, which is normally used in friction-loss calculations, is derived from the reciprocal of the 0.54 power, which is in the basic Hazen–Williams equation. The reciprocal of 0.54 is actually a repeating decimal, 1.85185185.... Near the beginning of the book, we discussed the fact that the reciprocal of 1.85 is a repeating decimal, 0.54054054...and pointed out that rounding off to 0.54 will occasionally yield a discrepancy. But now you have learned that 0.54 is the basic number and the rounding off errors should be blamed on 1.85, not 0.54. Not being a purist, we will suggest, somewhat arbitrarily, that we treat the commonly used form of the Hazen–Williams formula, Eq. 3, as the pure form and go from there. It should be understood that this formula is not derived; it is simply an empirical formula that has been shown to approximate what happens when freshwater flows through pipes at the temperatures, pressures, and turbulent flow rates normally experienced in fire protection systems.

Before going on, we will note that the Hazen–Williams equation, sometimes known as the Williams–Hazen equation, does not enjoy universal support outside of the fire protection field.

Some authorities have a strong preference for the Darcy–Weisbach equation and may go so far as to say that the Hazen–Williams equation displays an ignorance of basic turbulent flow. It has been said that at the onset of turbulent flow, the turbulence is limited to the core, with a laminar film hiding the roughness of the pipe. As the flow rate increases, this film gradually breaks down until at some point the pipe roughness is fully exposed. It has been said that at lower flow rates (perhaps up to 12 ft. per sec.), the friction loss may be overestimated by 20–45% when calculated by the Hazen–Williams equation. Of course, some of us who are concerned about the lack of a safety factor in many sprinkler system designs may be inclined to take what we can get when we get it. Let's hear it for Hazen–Williams (or should it be Williams–Hazen?)! Just always remember, though, the subject of this book is not an exact science.

This prompts me to digress on my digression. Many years ago, a Society of Fire Protection Engineers chapter, desperate for speakers, endured a few remarks from me. Along the way, I casually commented that fire protection was more an art than a science. I was surprised to find that a few members took strong exception. Since that time, there has been extensive research in many areas and great strides have been made with computer models that were unheard of in those days. Despite all the valuable knowledge that has been acquired, I am still prepared to argue that fire protection is more an art than a science. Every building is unique. Further, the multitudinous factors within the building that are relevant when there is a fire are constantly changing. Computer programs that attempt to tell you when the first sprinkler will activate or how many minutes you have to evacuate a building have value in terms of providing a better "point of departure" in analyzing a specific problem, but I will argue that they do not really move fire protection all the way from an art to a science. Now back to the subject at hand.

**Consider the "*C*"**

The Hazen–Williams coefficient "*C*" is a measure of the roughness of the interior wall of the pipe. You can see that the higher the value of the "*C*", the lower the friction loss.

For the purpose of calculating a sprinkler system, the following "*C*" values, which take into account the deterioration of the "*C*" factor over time, are prescribed by NFPA 13.

| | |
|---|---|
| Unlined Cast or Ductile Iron | 100 |
| Black Steel (Dry Systems, including Preaction) | 100 |
| Black Steel (Wet Systems, including Deluge) | 120 |
| Galvanized (Dry Systems, including Preaction) | 100 |
| Galvanized Steel (Wet systems, including Deluge) | 120 |
| Plastic (listed)—all | 150 |
| Cement-Lined Cast—or Ductile Iron | 140 |
| Copper Tube, Brass, or Stainless Steel | 150 |
| Asbestos cement | 140 |
| Concrete | 140 |

These "*C*" values are one of many approximations in a calculated system. The actual "*C*" value of new steel pipe is at least 140, for both wet and dry sprinkler systems. But testing of aged sprinkler systems has led the NFPA 13 Committee to specify the reduced values to better simulate long-term performance.

In the early 1980s, tests by what was then the National Bureau of Standards on new steel pipe yielded an average "*C*" of 148.3 and the claim was made that under the test conditions, the pipe was much more vulnerable to corrosion than normal pipe in service in a wet pipe system. An unsuccessful effort was made to change NFPA 13's requirement for a "*C*" of 120 in calculations for a wet pipe system to a "*C*" of 140.

The 140 value used for the usual underground pipe supplying the sprinkler system assumes that this is a true fire main that normally has no flow, and 140 should be used in hydraulic calculations only when that is the case. Tuberculation is a function of the flow and the properties of the water. Street mains have a decaying "*C*" value, which can even drop below 50 after a period of many years.

It is possible, by making hydrant-flow tests, to determine the "*C*" in a street main. Consider the following example:

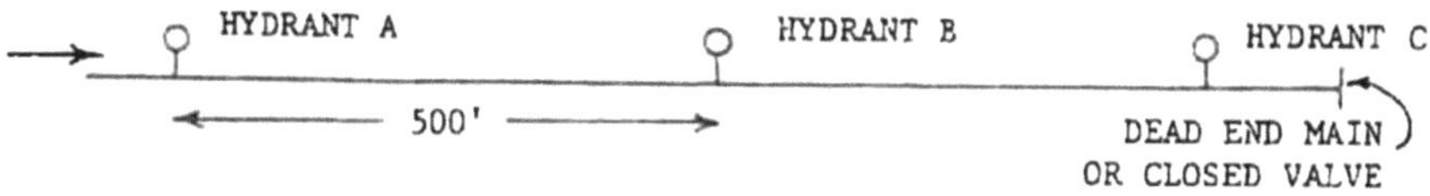


It is necessary to have one-way flow and all of the flow coming from upstream of Hydrant A. *Caution:* If valves must be closed to make this test, carefully evaluate the consequences of closing the valves before doing so, take any appropriate precautions, and take great care that all valves are restored to the full open position immediately following the test.

Flow Hydrant *C* and take static and residual readings at both Hydrant A and Hydrant B. Let us assume the following results:Flow at Hydrant C: 940 gpm

Pressure readings at Hydrant A: Static: 85, Residual: 57
Pressure readings at Hydrant B: Static: 82, Residual: 45

First, consider the difference of 3 psi in the two static pressures. Assuming accurately calibrated gauges, this can be explained by an elevation difference (about 7 feet). The significant number is the drop in pressure with the hydrant flowing (indicated by the residual pressure), which is 28 psi at Hydrant A and 37 psi at Hydrant B. This indicates that the friction loss between Hydrants A and B is 37 − 28 = 9 psi.

With a little algebraic manipulation, the Hazen–Williams formula becomes where $Q = 940$

$$C = \left(\frac{4.52}{pd^{4.87}}\right)^{0.54} \times Q \tag{4}$$

$$p(\mathit{friction\ loss, in\ psi, per\ foot\ of\ pipe}) = \frac{9}{500}$$

$d =$ 8 (The internal diameter of 8-inch underground pipe varies according to type. While it will never be exactly 8 in., this is close enough for purposes of this calculation.)

Thus,

$$C = \left( \frac{4.52}{\frac{9}{500} \times 8^{4.87}} \right)^{0.54} \times 940 \approx 78$$

## Consider the "*d*"

Like everything else, in the good old days, this was very simple. Schedule 40 pipe was mandated for sprinkler systems and the internal diameter of 2-in. pipe was always 2.067 in. Along the way, copper tube gained acceptance after overcoming concerns about joining methods that would withstand fire conditions, but it is not widely used. Schedule 10 pipe, also referred to as thin wall, made its appearance in the 1970s although minimum wall thickness requirements initially limited its use to sizes under 4 in., with a minimum wall thickness of 0.188 in. required for 4-in. and larger pipe. The 1978 Edition of NFPA 13 extended the use of Schedule 10 for sizes up through 5-inch. 6-inch. pipe must have a minimum wall thickness of 0.134 in. and a minimum wall thickness of 0.188 inches applies to 8-in. and 10-in. pipe.

As discussed beginning on page 47, there has long been wording permitting the development of more efficient or economical piping.

Plastic pipe was introduced in 1984 and is currently listed for Residential and Light Hazard Occupancies, subject to a number of restrictions which must be carefully adhered to. Refer to the tables in Appendix G for some of the types of listed piping currently available.

Sprinkler drawings sometimes fail to show the kind of pipe that is to be used and, in such instances, you must refer to the calculations to determine what kinds of pipe are contemplated. We say "contemplated" because who can be sure that what is installed is what was used in the calculations?

The 1994 Edition of NFPA 13 finally addressed the problem of the inspector who must determine what kind of pipe has been installed. It requires that "all pipe…be marked continuously along its length by the manufacturer" and "this identification shall include the manufacturer's name, model designation, or schedule."

**Back to the Overall Formula**:

For a given "*C*" value and pipe size, the Hazen–Williams formula can be reduced to a constant, which, when multiplied by any flow ($Q$) to the 1.85 power, will give the friction loss in psi per linear foot. Tables with these constants, which can be used if you have a calculator with the $y^x$ function, are in Table 4 of Appendix A and in Appendix G. There are also published friction-loss tables. One such table, for Schedule 40 pipe and a "*C*" value of 120, will be found in Appendix E, along with conversion factors that may be applied to this table.

# Underground Fire Service Mains

Underground fire service mains can be made of a variety of materials, including cast iron, ductile iron, steel, asbestos cement, plastics such as PVC, and fiber-reinforced composites. Modern cast-iron, ductile-iron, and steel pipe is normally cement lined.

A Hazen–Williams "*C*" of 150 can be used for plastic pipe. Unlined cast-iron pipe is considered to have a "*C*" of 100. All cement-lined pipe is considered to have a "*C*" of 140.

The internal diameters of the different types of pipe vary considerably and, of course, the internal diameter is raised to the 4.87 power in the Hazen–Williams equation, which greatly magnifies the effect of the different diameters. Nominal 8-in. asbestos cement pipe may have an internal diameter of 7.85 in., whereas nominal 8-in. Class 52 cement-lined ductile iron pipe may have an internal diameter of 8.265 in., which means that, although both pipes can be considered to have a "*C*" of 140, the friction loss in the asbestos cement pipe will be 28.5% greater than in the Class 52 ductile iron pipe. And we have not selected the extremes.

Many submissions of calculated sprinkler systems contain no information on the underground pipe. They usually do show an internal diameter for the underground in their calculations, and you are expected to accept it on faith. Generally, however, the friction loss in the underground in question does not exceed a few psi, making even a substantial percentage error fairly insignificant. For this reason, in the absence of specific information, we see no objection to simply plugging in the nominal pipe diameter. Only in the rare case where the loss is significant is it necessary to make an effort to ascertain the correct inside diameter. Refer to Table 4 in Appendix A for friction-loss constants based upon the nominal diameter and a Hazen–Williams "*C*" of 140.



H. S. Wass Jr. and R. P. Fleming P.E., *Sprinkler Hydraulics*,
https://doi.org/10.1007/978-3-030-02595-3_13

# Losses from Fittings and Valves

When water flowing through a pipe encounters an obstacle or changes direction, there is a head loss. This is cranked into the calculations in terms of EQUIVALENT FEET OF PIPE, which is added to the actual pipe length of the section for which the friction loss is being calculated. It should be noted that this loss is independent of the "*C*" value of the pipe. Theoretically, it is not independent of the rate of flow, but it is reasonable to ignore this fact for normal flow rates (see page 85).

Refer to Table 1 in Appendix A for suggested Equivalent Pipe Lengths (EPLs) for Schedule 40 steel pipe. Note carefully that there is an adjustment factor to be applied to the equivalent pipe lengths when the value of "*C*" is other than 120. The effect of this adjustment is to make the friction loss through the fitting the same for all *C*-values. The 1994 Edition of NFPA 13 introduced a second adjustment factor. Since then, when the internal diameter is different from Schedule 40 pipe, the equivalent feet in this table is also to be multiplied by the following factor:

$$Factor = \left(\frac{Actual\ inside\ diameter}{Schedule\ 40\ steel\ pipe\ inside\ diameter}\right)^{4.87}$$

The effect is, again, to generate the same friction loss, in psi, for the fitting as for a Schedule 40 system. The submitter of this change stated that "the loss through the same fittings with the same flow should be the same regardless of the internal diameter." Presumably, he meant regardless of the internal diameter of the pipe, not the fittings. He cites an example where the friction loss in a run of 3-in. pipe with three elbows and one tee flowing 600 gpm could be almost 5 psi less with a large internal diameter nominal 3-in. pipe. Interestingly, one jurisdiction (the Texas State Board of Insurance) called for this many years ago. As a practical matter, this change was not be reflected in all of the actual calculations being performed for some time since the commercial computer programs had to be revised to make this calculation.



H. S. Wass Jr. and R. P. Fleming P.E., *Sprinkler Hydraulics*,
https://doi.org/10.1007/978-3-030-02595-3_14

NFPA 13 also states that the equivalent pipe length table should be used "unless manufacturer's test data indicate that other factors are appropriate." The manufacturers of plastic pipe and copper tubing provide equivalent pipe length tables for their products. Appendix A contains suggested equivalent pipe lengths for plastic pipe and copper tubing. These tables should be used only if you do not have access to the manufacturer's data since their data may differ slightly in some instances.

NFPA 13 Section 27.2.4.8.1 provides the following rules for calculations through the system piping:

(1) Pipe, fittings, and devices such as valves, meters, flow switches in pipes 2-in. or less in size, and strainers shall be included, and elevation changes that affect the sprinkler discharge shall be calculated.
(2) Tie-in drain piping shall not be included in the hydraulic calculations.
(3) The loss for a tee or cross shall be calculated where flow direction change occurs based on the equivalent pipe length of the piping segment in which the fitting is included.
(4) The tee at the top of a riser nipple shall be included in the branch line, the tee at the base of a riser nipple shall be included in the riser nipple, and the tee or cross at a cross main-feed main junction shall be included in the cross main.
(5) Fitting loss for straight-through flow in a tee or cross shall not be included.
(6) The loss or reducing elbows based on the equivalent feet value of the smallest outlet shall be calculated.
(7) The equivalent feet value for the standard elbow on any abrupt 90° turn, such as the screw-type pattern, shall be used.
(8) The equivalent feet value for the long-turn elbow on any sweeping 90° turn, such as a flanged, welded, or mechanical joint—elbow type shall be used.
(9) Friction loss shall be excluded for the fitting directly attached to a sprinkler.
(10) Losses through a pressure-reducing valve shall be included based on the normal inlet pressure condition. Pressure loss data from the manufacturer's literature shall be used.

Fittings, particularly tees, which are commonly found at the beginning of branch lines, can have a very significant effect upon the hydraulics of a system, and care must be taken to include all of them in the calculation, with the exception noted above in subsection (9).

When NFPA 13 says that "friction loss shall be excluded for the fitting directly connected to a sprinkler," they are talking about the tee or, in the case of the end head on a branch line, the elbow to which the sprinkler is commonly attached. A reader of the FM *Handbook of Industrial Loss Prevention*, published in 1967, will find contrary advice where they specifically include the loss in the elbow to which the end-of-the-line head is attached. One also might wonder if the addition of a short nipple between the sprinkler and the tee on the branch lines should lead to the quantum change resulting from including the five-foot equivalent length of a one-inch tee.

There is, or was, a sprinkler manufactured with an attached adjustable (±3/4″) nipple 3-3/4-in. long for use with dropped ceilings. The manufacturer lists the "*k*" for this assembly as 5.53, whereas the "*k*" for the sprinkler, by itself, is listed as 5.62. You can make your own theoretical calculations on this. If you are not sure how to go about this, now may be a good time to discuss it.

Drop nipples to the individual sprinklers are used where there are suspended ceilings with the piping concealed above it. Riser nipples (going up instead of down) are occasionally encountered. To keep it simple, assume the normal case where all of the sprinklers are at the same level. Ignore the elevation change produced by the nipple until you reach the main riser, at which time you adjust for the elevation differences between the base of the riser and the sprinklers. When the nipple is part of an assembly with the sprinkler attached, as in the example cited above, simply use the "*k*" provided by the manufacturer; otherwise, it is necessary to calculate a "*k*" for the sprinkler-nipple assembly. We will explain this by an example.

Assume a two-foot-long nipple of one-inch nominal size attached to a tee in the branch line. Since we ignore elevation at this stage, it does not matter whether it goes up or down. The sprinkler attached to the nipple has a "*k*" of 5.65. Take a typical flow for the design, which we will assume to be 25.0 gpm, and calculate the associated pressure:

$$p = \left(\frac{Q}{k}\right)^2 = 19.58\,\text{psi}$$

Now calculate the friction loss at a flow of 25.0 gpm, from the sprinkler through the tee on the branch line. Two feet of actual pipe length plus 5 equivalent feet for the tee equals 7 feet. Multiplying by a friction loss per foot of 0.197 yields 1.38 psi, which is added to the pressure at the sprinkler, 19.58 to get 20.96 psi. We now can calculate the "*k*" for the assembly.

$$k = \frac{Q}{\sqrt{p}} = \frac{25.0}{\sqrt{20.96}} = 5.46$$

We are not sure how closely this relates to the "real-life" flow, but it conforms to NFPA 13, and we are not yet prepared to recommend anything better.

It is even more important to calculate a "*k*" when you have what NFPA calls a "return bend", also known as armovers.

As you can see, the reduction in the value of the "*k*" is even more significant since there are two elbows, or four equivalent feet, in addition to the tee.

Use the equivalent pipe length table with caution. There are many special fittings listed for specific types of pipe that simplify their installation but result in a higher-pressure loss than standard fittings. In some instances, the difference is substantial.

Those who have worked with flow through pipes in other fields may be looking for something to plug in where the pipe size changes. The 1976 Edition of NFPA 13 said that friction loss in reducers should be excluded. This statement was deleted in the next edition although nothing was added to include friction loss in reducers. Presumably, the loss is not significant.

NFPA 13 does not specifically address the fittings that are found in grids where the branch lines are typically connected at each end to a feed main with a tee. In general, the tees at each end of the branch lines are always included in the branch line. Except for the branch lines where sprinklers are assumed to be flowing, the flow is from a cross main into a smaller branch line on one end and from a branch line into a larger cross main on the other end. While the loss through the two tees is considered to be equal, one might wonder if this is really the case. Intuitively, you might suspect the loss would be less when flowing from a smaller pipe into a larger pipe. A brief search of general literature on flow in pipes suggests the opposite may be true. Of course, as we said at the outset, we are only dealing with approximations. Perhaps the difference is not sufficiently significant to warrant a further complication in calculating grids. We really don't know.

# Backflow Preventers

I once received a letter from a fire protection and code consultant with considerable experience with backflow preventers who offered the opinion that "they pose the single biggest threat to operating fire sprinkler systems since glued-in sprinkler heads." The catalog of one manufacturer of backflow preventers takes a more benign view. "The fire sprinkler community's increased awareness of the need for backflow prevention protection and stronger regulations requiring protection have helped to establish a need for backflow preventers in fire protection systems." In reading this quotation, give more credence to "stronger regulations" than "increased awareness." That is not to suggest that the fire protection community is either ignorant or irresponsible. Rather, there are issues of cost and, more importantly, friction loss and reliability, that the fire sprinkler community weighs along with the many years of experience with just a single check valve, which does not constitute "backflow prevention" as currently defined.

Backflow prevention is not a simple matter. One insurance company understated the case when they said, "the public expects the water supplied by public utilities to be safe to drink." Of course we do. But, as with all safety and health issues, 100% certainty is unattainable. Honest differences of opinion are inevitable when balancing cost versus benefit.

The source of the standards for public water supplies is the American Water Works Association. As they have acknowledged, as our society advances, we set higher standards. To quote from an old AWWA publication:

> Years ago the water purveyor was satisfied if the quality of water he distributed to his customers met the following standard: The poor could use it for making soup; the middle class, for dyeing their clothes; and the rich for watering their lawns. If the people who drank the water filtered it through a ladder, disinfected it with chloride of lime, and then lifted out the dangerous germs that survived and killed them with a club, the water was considered fit to drink.

The writer then points out that "today's water utility must supply water to its customers that is not only safe, but free of objectionable taste and odor, color, turbidity, and staining elements."



H. S. Wass Jr. and R. P. Fleming P.E., *Sprinkler Hydraulics*,
https://doi.org/10.1007/978-3-030-02595-3_15

This brings us to what is known in the trade as "cross-connections." American Water Works Association Manual M 14, Recommended Practice for Backflow Prevention and Cross-Connection Control, has defined a cross-connection as "any connection or structural arrangement between a public or a consumer's potable water system and any nonpotable source or system through which backflow can occur." Amplifying upon this, they point out that there are direct and indirect cross-connections. A direct connection means a physical joining between "safe" and "unsafe" water. An indirect connection "is an arrangement whereby unsafe water in a system may be blown, sucked, or otherwise diverted into a safe water system." As an example of an indirect connection, they mention a lavatory wash-basin where the faucet extends downward below the level of the water that might accumulate if the stopper were closed. Of course, our civilization has advanced to the point where you will not encounter this; just one of the subtle little niceties we are not even aware of.

Having mentioned AWWA Manual 14, this is a good place to amplify upon it. Unlike NFPA 13, Manual 14 is just a manual, not a standard. The significance of this distinction lies in the lack of an opportunity for public review and input. Manual 14 establishes six classes of fire protection systems, based upon the water sources and their arrangement.

What they call a Class 1 system is the common direct connection to a public water main with no other water supplies and no antifreeze or other additives in the sprinkler lines. A Class 2 system is a Class 1 system with a booster pump. The other classes get into tanks, reservoirs, secondary sources of water, etc., where more concerns can arise.

When M 14 was revised in 1990 (for the first time since 1966), one significant new item said that "where the fire sprinkler piping is not an acceptable potable water system material, there shall be a backflow prevention assembly isolating the fire sprinkler system from the potable water system." Since black steel pipe is not considered a potable water pipe, for the first time, the AWWA was saying that a single check valve is not sufficient for a sprinkler supply main. In and of itself, the American Water Works Association has no authority but their words filter down to the code-making groups that write plumbing codes.

Historically, the jurisdiction of NFPA 13 stopped at the connection to the public water supply. NFPA 24, Installation of Private Fire Service Mains and their Appurtenances, took over at this point. Along the way, NFPA 24 wisely felt called upon to acknowledge the contamination issue in Classes 3–6 systems with a non-mandatory appendix statement that an approved reduced-pressure-zone-type backflow preventer is "recommended" if there is an antifreeze system, an auxiliary water supply interconnected with the public supply or an auxiliary water supply for fire department use. If there is an auxiliary supply from a tank or reservoir maintained in potable condition, an approved double check valve assembly is "recommended." The 1999 Edition of NFPA 13, which incorporated much of NFPA 24, included this appendix item.

Backflow, the big concern, has two causes:

1. Backsiphonage. This occurs when the pressure in the "good" water in the public water supply becomes less than atmospheric pressure, creating the potential to pull objectionable water into the public supply.
2. Backpressure. This occurs when the "good" water is minding its own business (staying at normal supply pressures) but the pressure in the "bad" water exceeds the pressure in the "good" water.

Most automatic sprinkler systems are connected to a public water supply. Additionally, fire department pumper connections are an important element in an automatic sprinkler system and who knows what kind of stuff is coming out of those fire department pumpers?

Unfortunate cross-connections have occurred in modem times. Holy Cross College in Worcester, Massachusetts has fielded some fine football teams but, curiously, their football teams have other claims to fame. The tragic Coconut Grove fire in Boston on November 28, 1942 claimed 492 lives, including victims who might not have been at this nightclub had they not been celebrating a victory by the Holy Cross football team over a supposedly invincible Boston College, on anyone's list of great sports upsets. In 1969, the entire football team was stricken with infectious hepatitis and the remainder of the season canceled. A cross-connection (not related to a sprinkler system) took the blame.

There does not seem to be any documented evidence that, in their some 120 years of existence, the water from an automatic sprinkler system has so much as slowed up a chess game (not that anyone would have noticed) let alone caused the cancelation of a football game. But I think we all are aware of the world we are living in, including the need to support lawyers in the manner to which they have been educated. We have read somewhere in the literature that an inadequately safeguarded cross-connection to a sprinkler system may have caused one death in this century. We have no way of knowing how many lives have been saved by sprinkler systems but it is a substantial number.

In recent years, even before the 1990 revision to Manual M 14, more and more jurisdictions have been mandating backflow preventers for automatic sprinkler systems. "Backflow prevention" devices usually mean one of two kinds of hardware:

1. Double Check Valve Assembly (DCVA).
2. Reduced-Pressure-Zone principle backflow prevention assembly (RPZ).

A double check valve assembly consists of two independently acting spring-loaded check valves spaced sufficiently that a foreign object should only impede one of them. A shutoff valve is provided at each end of the assembly to permit easy access for maintenance and test cocks are provided for checking the tightness of the check valves.

A reduced-pressure-zone backflow preventer, commonly referred to as an RPZ device, also contains two check valves. The added feature is a pressure differential relief valve connected to the line between the two check valves and also connected, on the other side of a diaphragm, to the supply side of the upstream check valve. The pressure differential relief valve is designed to maintain a pressure in the line between the two check valves somewhat less than the pressure on the supply side. Thus, in the event of a backflow condition, the relief valve will discharge from its relief port whatever amount of water is necessary to maintain the lower pressure.

The RPZ device, developed in the early 1940s, is considered more reliable than the double check valve assembly in preventing backflow. The reliability of both types from our fire protection perspective, however, leaves much to be desired. Typically, the clappers travel along a guide or have lever mechanisms causing the clappers to move perpendicular to the seal faces. Unlike the old-fashioned simple swing check valve, these check valves are somewhat sophisticated. To quote from one manufacturer's literature:

> In normal operation, the independent, spring-loaded check valves remain closed until there is a demand for water. Each of the two check valves in series is designed to open at one psi pressure differential in the direction of flow. In the event pressure increases downstream of the unit, tending to reverse direction of flow, both check valves are closed to prevent backflow. If the second check valve is prevented from closing tightly, the first check valve will still provide protection from a backflow condition.

We at IRM encountered an 18-month-old RPZ installation where a small deposit buildup on a clapper guide prevented the clapper from reaching the fully open position. With the RPZ devices, there is also concern that the pressure relief valve could hang up in the open position, effectively diverting the water from the sprinkler system.

Backflow preventers require annual testing and maintenance with respect to their backflow function. The testing normally must be performed by certified personnel and involves an operational testing of the pressure differential relief valve and tightness testing of the two check valves. The tests *do not* verify that the device will provide full flow under fire conditions.

NFPA 25, which sets forth the "inspection, testing, and maintenance requirements" from the fire protection perspective, has the following to say:

All backflow devices...shall be tested annually at the designed flow rate of the sprinkler system, including hose stream demands if appropriate, and the friction loss across the device measured and compared to the device manufacturer's specifications.

An exception is provided that where "connections of sufficient size to conduct a full flow test are not available, tests shall be conducted at the maximum flow rate possible."

After initially taking the view that a full flow test downstream of the backflow preventer was not necessary, that flow through the drain line would sufficiently "exercise" the internal springs in backflow devices, further discussion in the NFPA 13 committee led to a requirement for means for "full flow tests at system demand," which first appeared in the 1996 Edition.

Outside Stem and Yoke (OS&Y) valves isolating the backflow preventer should be inspected monthly if the valves are locked or electrically supervised; weekly otherwise. Incidentally, backflow preventers are normally available with both indicating (OS&Y) valves and non-indicating valves, with the OS&Ys costing more. OS&Y valves should always be provided on fire protection installations but occasionally this is not done.

Additional requirements for an RPZ device are

1. Weekly inspection "to ensure the differential sensing valve relief port is not continuously discharging." As pointed out in the Appendix, intermittent discharge is normal but continuous discharge indicates "fouling" of either or both of the check valves.
2. Inspection after any testing or repair to ensure that the valves have been properly restored in the open position. This is another problem we have encountered. The testing of the backflow function is not performed by people oriented toward fire protection. They may be more casual about restoring the system since with other kinds of water supplies, failure to restore the valves will be promptly detected by the user.

Where a backflow preventer is installed, only a listed or approved device should be used. UL says that "these devices have been classified as to friction loss and body strength only. Features for use in potable water systems have not been evaluated." FM Approvals advises that "the backflow preventers have been evaluated for reliability from a fire protection standpoint. No attempt has been made to determine their suitability to ensure the public health." Unlike check valves listed for fire service use, there is no functional test of the clapper assemblies on a backflow preventer.

Manual M 14 tells us that the approving agency for a backflow preventer (looking at the public health, not the fire protection aspect) is the University of Southern California Foundation for Cross-Connection Control and Hydraulic Research. Badly in need of abbreviation, this is referred to as the FCCCHR, which is probably about the best that can be done. Reportedly, their testing is limited to a pressure test followed by a 12-month field evaluation at sites selected by the manufacturer. Despite the theoretical soundness of the design of these devices, there is some evidence that, in practice, they may not always perform properly in terms of preventing backflow.

As has been stated, the RPZ device occasionally discharges water and can discharge at a high rate under adverse conditions. For this reason, it cannot be installed in a pit. Therefore, the preferred location for the device is outdoors above ground. In most parts of the country, where freezing conditions must be anticipated, it must be located inside of the building or in a heated outside enclosure. In all cases, adequate provision must be made for drainage and an air gap must be provided.

RPZ devices can add significantly to the cost of a sprinkler installation. The cost of installation and providing for adequate drainage add to the base cost of the equipment.

Now let's look at why so much space is being devoted to RPZ devices in a book on sprinkler hydraulics. You guessed it: friction loss. The friction loss varies with the make, model, size, and the rate of flow and can be significant, since internal check valves are loaded to close tightly. The manufacturers provide friction loss charts in their literature but we have been told that these charts are not always reliable. One manufacturer's literature heads the charts with "Flow Curves as established by the USC Foundation for Cross-Connection Control and Hydraulics Research," seeming to imply that the FCCCHR is the source. It is our understanding that the FCCCHR does not perform these tests. Underwriters Laboratories and Factory Mutual provide friction loss data for a limited number of flow rates, requiring some kind of crude interpolation for the rate of flow imposed by the sprinkler demand. Underwriters Laboratories published a standard, UL 1469, "Strength of Body and Hydraulic Pressure Loss Testing of Backflow Special Check Valves" which became effective on May 1, 1996. Aside from more rigorous testing of backflow prevention devices intended for fire protection use, extensive pressure loss data is established over the range of 0–100% of the rated flow.

The problems should be apparent:

1. It is vitally important to include the friction loss in the sprinkler calculations when there is an RPZ device in the supply line. This requires verifying whether or not there is such a device and, when there is one, estimating the friction loss at the calculated sprinkler flow rate.
2. An RPZ device must be inspected and maintained. When practical, an annual flow test at a rate approximating the sprinkler demand (plus hose demand when applicable) should be made to determine the friction loss through the backflow preventer. At the very least, two-inch drain tests should be performed regularly and carefully evaluated. It was a two-inch drain test that revealed the problem at the IRM location mentioned previously.
3. Quite aside from the substantial cost to install an RPZ device, the reduced pressure resulting from the friction loss through the device might dictate a larger diameter pipe, another added cost. A booster pump could be required, resulting not only in further expenses but reduced reliability (as discussed elsewhere).
4. In some jurisdictions, authorities have been mandating retrofits of existing installations. It can only be hoped that the possible consequences to the sprinkler protection are recognized and evaluated, with corrective measures being taken wherever needed. Since hydraulically designed sprinkler systems are typically calculated very close to the water supply, system reinforcement will normally be required.

Where a double check valve assembly is required, rather than an RPZ device, the retrofit concerns still apply but the overall impact is less. As mentioned earlier, a double check valve assembly is more likely to function properly when called upon to handle a full fire flow. Also, while the friction loss through the assembly is significant it is substantially less than through an RPZ device. The cost is also less.

And, of course, there is no need to provide for drainage and the installation can be in a conventional pit.

While the friction loss in the double check valve assembly is less than the RPZ devices, it is still significant enough that it is important for it to be included in the sprinkler system hydraulic calculations. And, again, when there is a retrofit, some kind of reinforcement will probably be required in the typical design which takes full advantage of the water supply.

Following are typical charts provided by one manufacturer showing friction loss in their devices:

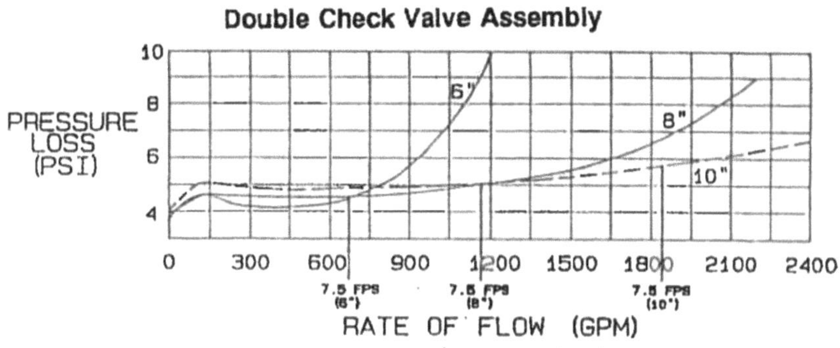


(Reprinted with permission of ConBraco Industries, Inc.)

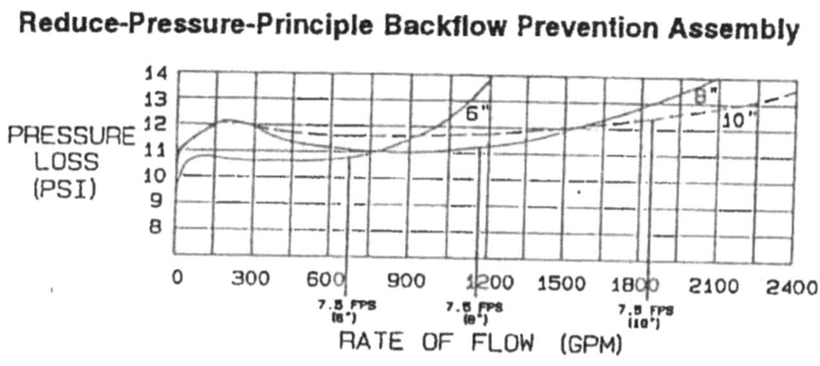


(Reprinted with permission by ConBraco Industries, Inc.)

It is clearly evident that the friction loss through the device is not a direct function of the flow rate. With a standard swing check valve, we can plug in a number of equivalent feet and get a reasonable approximation of the loss through the valve at any flow rate. There is no simple method to enter the loss from backflow devices into the calculations.

We will add one final note. The June 1999 issue of *Plumbing Engineer* had an article entitled "Backflow Prevention for Fire Protection Systems" written by a manufacturer's representative (presumably a manufacturer of backflow devices). He referred to an independent study released by the American Water Works Association Research Foundation entitled *The Impact of Wet-Piping Fire Sprinkler Systems on Drinking Water Quality*. He stated that "the study concluded that backflow does occur and further, that a health hazard does exist from water in fire sprinkler standpipes." He added, however, that "further research was recommended," suggesting to me no firm conclusions as to the real-world hazard although he headed this portion of his article with the phrase "study confirms need." Anyone wishing to go beyond my glib comments should contact the American Water Works Association.

# Velocity Pressure

Daniel Bernoulli (1700–1782), a member of a famous family of Swiss mathematicians and physicists, published *Hydrodynamica* in 1738 and gained recognition for applying Newton's law of the conservation of energy to the flow of liquids. For a frictionless, incompressible fluid, this is usually expressed in the following terms:

$$\frac{p_1}{w} + \frac{v_1^2}{2g} + z_1 = \frac{p_2}{w} + \frac{v_2^2}{2g} + z_2$$

Although this expression does not appear in Bernoulli's works, it is commonly known as Bernoulli's equation, Bernoulli's principle, or Bernoulli's theorem.

$p$ pressure, pounds per square foot
$w$ weight of water, pounds per cubic foot (62.4)
$v$ velocity, feet per second
$g$ acceleration of gravity, feet per second per second (32.2)
$z$ elevation head, feet
$\frac{p}{w}$ pressure head, feet, acting perpendicular to the pipe wall
$\frac{v^2}{2g}$ velocity head, feet, acting parallel to the pipe wall

Bernoulli's equation states that the sum of the pressure head, the velocity head, and the elevation head remains the same throughout a closed system, in the absence of friction. While for practical purposes, water can be considered an incompressible fluid, it cannot be considered frictionless, and a fourth element, $h_2$, representing the friction loss from 1 to 2 must be added to the right side of the equation.

The total pressure at the last flowing sprinkler on a line is translated into flow, and this is the pressure used in the discharge formula, $Q = k\sqrt{p}$. For all other flowing heads in the line, the "$p$" in the discharge formula should not include the velocity pressure since the velocity pressure acts parallel to the pipe wall. The total



H. S. Wass Jr. and R. P. Fleming P.E., *Sprinkler Hydraulics*,
https://doi.org/10.1007/978-3-030-02595-3_16

pressure minus the velocity pressure is known as the normal pressure, and it is the normal pressure that should be used in the discharge formula for other than the last flowing head in a line.

$$P_T = P_N + P_v$$

Thus,

$$P_N = P_T - P_v \tag{5}$$

To calculate velocity pressure, it is necessary to convert the velocity head,

$$\frac{v^2}{2g}$$

in Bernoulli's equation, in feet, into appropriate terms, psi. Since velocity pressure in psi is 62.4/144 (or 0.4333…) × velocity head, velocity pressure can be expressed as 62.4/144 × $v^2/2g$

$$\text{Intuitively (?)}, \ Q = av$$

where

$a$ is the cross-sectional area of the pipe, in sq. ft., and
$v$ is the flow velocity in ft./sec

Therefore, $v = \frac{Q}{a}$

Since velocity is expressed in feet per second, $Q$ must be expressed in cubic feet per second. We use gallons per minute in sprinkler calculations. Therefore, we must convert gallons per minute to cubic feet per second.

Thus

$$Q\left(\frac{\text{ft.}^3}{\text{sec}}\right) = \frac{Q(\text{gpm})}{60 \times 7.4805}$$

$$a = \frac{\pi D^2}{4}$$

but since the pipe diameter is normally expressed m inches, $D$ must be divided by 12 to convert from inches to feet.

Thus

$$a = \frac{\pi\left(\frac{D}{12}\right)^2}{4}$$

Substituting,

$$v = \frac{Q}{a} = \frac{\frac{Q}{60\times 7.4805}}{\frac{\pi\left(\frac{D}{12}\right)^2}{4}} = \frac{4 \times 144Q}{60 \times 7.4805 \times \pi D^2} = \frac{576Q}{448.83\pi D^2}$$

And

$$p_v = \frac{62.4}{144} \times \frac{v^2}{2g} = \frac{62.4}{144} \times \frac{\left[\frac{576Q}{448.83\pi D^2}\right]^2}{2} \times 32.2 = 0.001123\frac{Q^2}{D^4}$$

Thus, velocity pressure is calculated by the formula

$$p_v = 0.001123\frac{Q^2}{D^4} \qquad (6)$$

where

$P$ = velocity pressure, in psi
$Q$ = flow, in gpm
$D$ = internal diameter of the pipe, in inches

0.001123 is the number prescribed by NFPA 13, which in 1978 broke its silence on this subject. This differed slightly from the $Q^2/888D^4$ that was found both in the NFPA Handbook and the Factory Mutual Handbook at that time, since 1/0.001123 = 890.47. But, based on the calculations we have just made, 0.001123 looks pretty good.

We used reasonable raw numbers, 7.4805 is the number of gallons in a cubic foot, 62.4 is the weight of one cubic foot of water at about 53 °F, 32.2 is "*g*" (the acceleration of gravity) in feet per second per second. 32.2 is the commonly used value of "*g*", but, strictly speaking, "*g*" varies according to the latitude and is slightly affected by the altitude and other factors. Near the equator, it might be in the vicinity of 32.1, which would yield the "888." In the parts of the world we are usually concerned with, 32.1 is too low, but 32.2 is a little on the high side. 32.174 is the preferred number, chosen because it is the average acceleration of gravity at sea level at 45° latitude. Thus we could make a case for a "g" that would yield 0 instead of NFPA 13's 0.001123, but that is quibbling.

Velocity pressure becomes a significant factor only at relatively high rates of flow in a pipe. Failure to take velocity pressure into account normally results in an error on the safe side. For this reason, the use of velocity pressure in calculations is optional per NFPA 13. Velocity pressure is frequently ignored in grid calculations. In the interest of simplicity and conservatism, an argument can be made for prohibiting the use of velocity pressure in all calculations. Not everyone agrees. Some have expressed the view that good engineering practice dictates that the calculations should reflect what really happens. They are right in theory but, considering the various kinds of approximations and minor errors found in the typical set of calculations being produced today, it can be helpful to have at least one error always on the conservative side.

NFPA 15, the Standard for Water Spray Fixed Systems for Fire Protection, has for many years suggested in its examples of calculations in Appendix A that when the velocity pressure is less than 5% of the total pressure, it is reasonable to ignore velocity pressure. The 1996 Edition of NFPA 15, for the first time, said that "correction for velocity pressure shall be included in the calculations" except when "the velocity pressure does not exceed 5% of the total pressure at each junction point." In the "Report on Proposals", the committee had this to say in their "rationale behind the substantive changes":

> "Neither NFPA 13-1994 nor NFP A 15-1990 provides guidance on when correction for velocity head should be made. If balancing at hydraulic junction points is to be required per NFPA 13 and 15 whenever there is a difference of 0.50 psi, the same logic would dictate that velocity head correction is necessary whenever the velocity exceeds 8.63 ft./s. Velocities in excess of 8.63 ft./s are common in water spray systems.
>
> Velocity pressure has the effect of reducing the flow from the side outlet of a junction. In a water spray system, all end nozzle pressure requirements must be met. Ignoring velocity head can introduce a significant error, resulting in an actual nozzle pressure that is less than required.
>
> Since the calculation method is tedious, the 'exception' allows some latitude. A difference of 5% pressure limits the error in flow rate to less than 3%."

We plead ignorance on the fine points of water spray systems and have no reason to question their concern. The reader may wonder how they arrived at the velocity of 8.63 ft./s. A little arithmetic reveals that 8.63 ft./s translates into a flow of about 23.25 gpm in one-inch Schedule 40 pipe and the velocity pressure with a flow of 23.25 gpm in Schedule 40 pipe is about 0.50. How they moved from there to a continuation of the magic "5%" is less clear but perhaps is simply motivated by the laudable desire not to make things too complicated.

The challenge to NFPA 13 to follow the dictate of "logic" does have some logic but we will stand by our previous comments.

NFPA 13 requires that "if velocity pressures are used, they shall be used on both branch lines and cross mains where applicable."

If velocity pressure is included in grid calculations, the sprinklers where the flow splits should be considered as end sprinklers, that is, all pressure is assumed to be translated into flow. See the section on the grid for a schematic showing what we mean by "the sprinklers where the flow splits." Likewise, with a loop supplying dead-end branch lines, no velocity pressure should be assumed for a branch line where a flow split occurs (although velocity pressure would be taken into account in the usual way for the individual sprinklers on that branch line). Again, no velocity pressure is involved for the last flowing branch line in a dead-end system. Velocity pressure in all other branch lines is calculated by treating the flow into the branch line in the same manner as a sprinkler at the junction of the branch line with the cross main.

Velocity pressure can be readily calculated by multiplying $Q^2$ (the flow in gpm, squared) by the appropriate constant. Refer to Table 5, Appendix A.

Inclusion of velocity pressure in the calculations involves a trial-and-error procedure. Following is an example that illustrates the process:

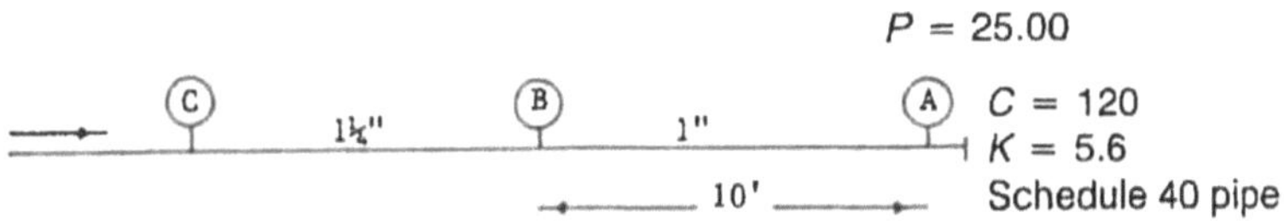


Assume that the pressure is 25 psi at sprinkler A. Velocity pressure is not a factor here because it is the end sprinkler flowing. The flow at sprinkler A is $k\sqrt{p} = 5.6\sqrt{25} = 28.0\,\text{gpm}$. With 28 gpm flowing in a 1″ pipe, the friction loss is 0.243 psi per foot. Multiplying by 10’, we determine that the friction loss between sprinkler A and sprinkler B is 2.43 psi. Adding this to 25 psi, the pressure at the end sprinkler, the *total pressure* at sprinkler B becomes 27.43 psi. Now deduct the *velocity pressure* to arrive at the *normal pressure*, which will be used to determine the flow from sprinkler B.

The velocity pressure at sprinkler B is a function of the rate of flow *approaching* sprinkler B, which is the combined flow of sprinklers A and B, and the size of the upstream pipe (1¼″). Since we cannot yet determine the flow from sprinkler B, it is necessary to *assume* a flow and make a *trial*. Let us assume a flow of 30 gpm from sprinkler B, which means a flow of 58 gpm between sprinkler C and B.

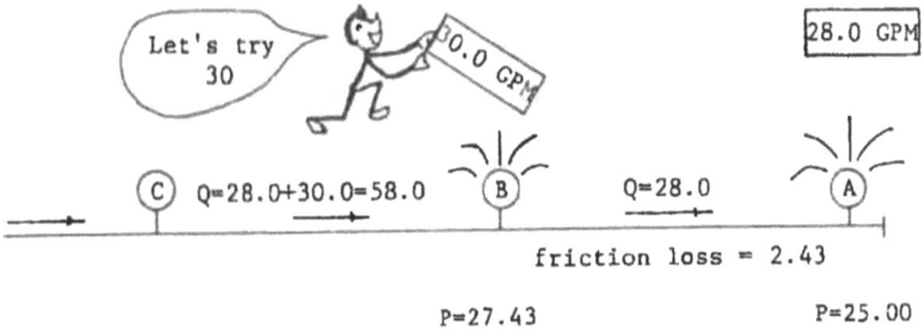


Remembering that we are concerned with the pipe on the supply side of sprinkler B, obtain the velocity pressure factor for the 1¼″ pipe from Table 5, Appendix A, 3.10 × 104, We now multiply this factor by the flow squared:

$3.10 \times 10^{-4} \times 58^2 = 1.04$ psi. Deducting 1.04 psi from the total pressure of 27.43 at sprinkler B, we get a normal pressure of 26.39 psi. This pressure results in a flow of $5.6\sqrt{26.39} = 28.8\,\text{gpm}$. This is lower than our assumed flow of 30 gpm from sprinkler B. Let us make a new trial, using the 28.8 gpm calculated on the first trial.

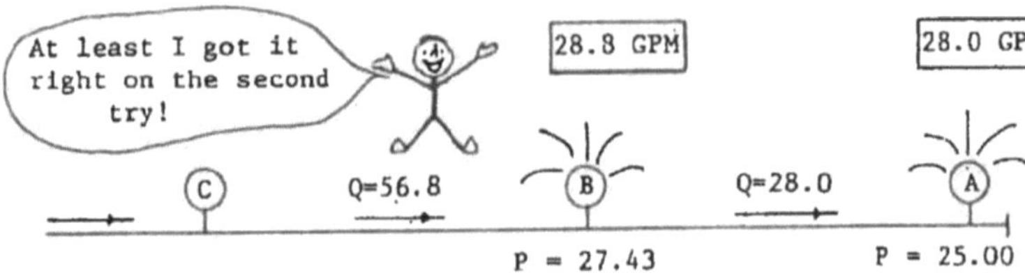


Our total flow becomes 56.8 gpm and $3.10 \times 10^{-4} \times 56.8^2 = 1.00$ psi. Deducting this from the total pressure of 27.43 produces a normal pressure of 26.43 and $Q = 5.6\sqrt{26.43} = 28.8\,\text{gpm}$, which is what we assumed in this second trial. This is the correct flow.

You will note that if the initial assumed flow is not excessively far off (in this case, the assumed flow was 30.0 gpm and the actual flow was 28.8 gpm), the flow computed on the first trial is the right answer. Thus, you usually should use that figure for the second trial, as we did in this example.

Note that if velocity pressure is ignored, the calculated flow at sprinkler B would be $Q = 5.6\sqrt{27.43} = 29.3\,\text{gpm}$, or 0.5 gpm higher than the "correct" flow. In tum, this higher flow would lead to greater calculated friction loss between sprinkler B and sprinkler C, which, in tum, would produce a higher calculated flow from sprinkler C. This effect would continue through the entire calculation, resulting in a higher total flow and pressure. Thus, the conservative result when velocity pressure is ignored.

# The Hydraulically Most Remote Area

The calculations normally must be made for the hydraulically most remote area. By hydraulically most remote, we mean the least favorable area; that is, when all flows from discharging sprinklers within the operating area are hydraulically calculated back to the source and the total friction loss is at a maximum. With a dead-end system ("tree" system), the hydraulically most remote area is usually obvious; with a grid, it is seldom obvious. With both types of systems, there are some pitfalls that may be encountered, and NFPA 13 has a few things to say on the subject. We will discuss dead-end systems and grids separately. First, however, we will digress to amplify on the design area, then we will look at a few pitfalls that may be encountered in either type of system and review a little history.

Most commonly the design area is derived using the "Density/Area" method using the appropriate graph or table in NFPA 13, and that is where the hydraulically remote area comes into play. Almost from the beginning, however, NPFA 13 has offered a second method for determining the design area, the "Room Design Method", which we discussed in the section "What Are We Calculating?." With the "Room Design Method," you must determine the hydraulically most *demanding* area, which relates to room size as well as remoteness from the riser.

While many systems maintain the same distance between sprinklers along the branch lines and the same distance between branch lines throughout, occasionally there will be different spacing within the same system. Assuming that the same density is required in all areas, you must keep in mind that the required flow per sprinkler increases as the spacing increases. Thus, if the area per sprinkler is greater in an area that would not otherwise be considered the hydraulically remote area, it may be necessary to make additional calculations to verify that the required density is met in this area of greater spacing. Occasionally, you encounter a system with no symmetry at all, such as a nursing home with an irregular shape and many small rooms. Even if the room design method has not been used, it is necessary to exercise careful judgment. When there is doubt, additional calculations should be made.

 
H. S. Wass Jr. and R. P. Fleming P.E., *Sprinkler Hydraulics*,
https://doi.org/10.1007/978-3-030-02595-3_17

NFPA 13 has a few special rules which make a certain amount of sense but, at the same time, increase the need for careful evaluation and judgment regarding the hydraulically remote area They say that “small compartments”, such as closets and washrooms, requiring only one sprinkler may be omitted from “hydraulic calculations within the area of application.” If the designer exercises this option, we suggest that calculations should be made for the most remote area that does not have these small rooms. This second set of calculations could yield a lower pressure but a higher flow at the base of the riser. Which is more “hydraulically remote”? It depends upon the water supply. It is necessary to plot the results on graph paper, as explained in the section on “Relating Hydraulic Calculations to the Water Supply.” Before leaving this subject, it should be noted that they stipulate that the sprinklers in small compartments omitted from the calculations shall, however, be capable of discharging the required minimum density. This seems reasonable enough until you start thinking about how you determine this when you are omitting these sprinklers from your calculations.

Although infrequent, and usually not recommended, there are times when different design criteria are established for different parts of the system. A rack storage area, for example, might be located near the riser. In such a case, separate calculations are necessary for the hydraulically remote area in that part of the system.

Aside from the location of the hydraulically remote area, a very important question is the shape of the area. The rules of the game with respect to the shape have changed over the years. Since you may encounter systems designed under the old rules, a brief history is in order.

Prior to the 1978 Edition, NFPA 13 said that, for dead-end systems, the area “usually includes sprinklers on both sides of the cross main” and the Appendix of NFPA 13 contained examples where they showed all heads on the physically remote branch lines included in the design area. With a center feed, many sprinklers on the branch lines, and a small design area, this could result in a long, narrow rectangle. Perhaps the word “usually” was inserted to permit discretion when the rectangle became very long and very narrow, but no guidance was provided.

Up through the 1976 Edition, NFPA 13 said that “for gridded systems, the design area shall be the hydraulically most remote area which approaches a square.” The defense of the square, of course, is that sprinklers tend to open over a circular area and a square can be viewed as a rough approximation of a circle. Since this argument applies without regard to the type of sprinkler system, and since the square is less hydraulically demanding than the long narrow rectangle that seemed to be required for the dead-end system, the prescribed treatment of the two types of systems was clearly inconsistent.

Common sense suggests that the square is too liberal and the long, narrow rectangle too conservative. While sprinklers tend to open over a circular area in a fire, that will not always be true. Under some circumstances, fire could spread rapidly along a row of storage or a conveyor belt. There could be strong draft conditions that could carry the fire and heat in one direction. A fire adjacent to a wall obviously will not open heads in a circular pattern.

As happens more often than the cynics among us think, common sense eventually prevailed. A Tentative Interim Amendment to NFPA 13 was issued on June 22, 1977, specifying that for a gridded system, the design area should have a dimension parallel to the branch line equal to 1.2 times the square root of the area. Whether or not this elongation of the design area along the branch lines is sufficient is, we think, an open question. It is an important question because as more sprinklers open along a branch line in a grid than were taken into account in the design area, there is usually a rapid decay in the pressure (and consequently, the discharge) because of the small pipe. Following the adoption of the "1.2 times the square root of the area" rule, several major insurance companies began calling for a factor of 1.4.

All discrimination against dead-end systems ended with the 1978 Edition of NFPA 13 when the Tentative Interim Amendment was incorporated with no substantive changes and extended to all systems. The wording was as follows:

> For all systems, the design area shall be the hydraulically most remote rectangular area having a dimension parallel to the branch line equal to, or greater than, 1.2 times the square root of the area of sprinkler operation corresponding to the density used. Any fractional sprinkler shall be carried to the next higher whole sprinkler.
>
> For gridded systems, the designer shall verify he is using the hydraulically most demanding area. A minimum of two additional sets of calculations shall be submitted to demonstrate peaking of remote area friction loss when compared to areas immediately adjacent on either side along the same branch lines.

An exception was added in 1980 to cover systems where the branch lines have an insufficient number of sprinklers to fulfill the $1.2\sqrt{A}$ requirement. Additional sprinklers on adjacent branch lines supplied by the same cross main must be included. The current wording has changed but the meaning is the same except that the designer is no longer assumed to be a "he."

We will illustrate the application of this rule. Assume a design area of 1500 sq. ft. The square root of 1500 is 38.7. Multiplying by 1.2 yields 46.4 (or 46.5 with less rounding off of the first step). If the sprinkler spacing along the branch line is 12.5 ft., 46.4 divided by 12.5 equals approximately 3.7, which means that four sprinklers are required on each branch line. If the spacing between the branch lines is 10 ft., the area per sprinkler is 10 × 12.5 sq. ft., and dividing 1500 by 125, we find that 12 sprinklers are required in the design area, or 4 sprinklers on each of the 3 most remote branch lines.

If there are more than four sprinklers on each branch line, the sprinklers selected would obviously be the four end sprinklers on the line. If there are only three sprinklers on each side of the cross main, you should select the three sprinklers on one of the branch lines and the adjacent sprinkler on the other side of the cross main.

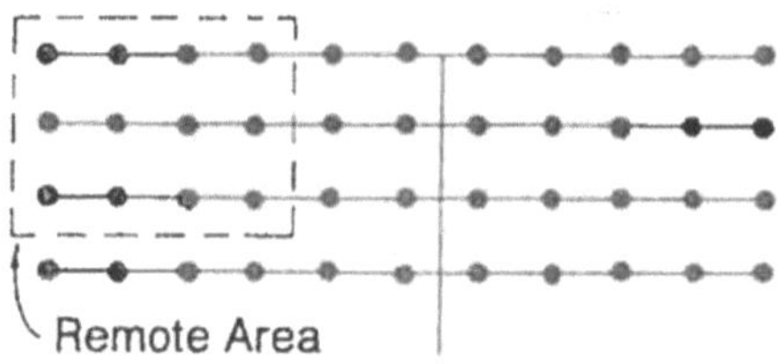


Design Area: 1500 sq. ft.

Distance Between sprinklers: 12.5 ft.

Distance Between Branch Lines: 10 ft.

Design Area:1500 sq.ft.

Distance Between sprinklers: 12.5 ft.

Distance Between Branch Lines:10 ft.

In this example, if the sprinkler spacing was 12 ft., rather than 12.5 ft., the area per sprinkler would be 120 sq. ft., which, divided by 1500, yields 12.5 meaning that 13 sprinklers are required in the design area. The extra sprinkler would be placed on the fourth branch line from the end, adjacent to any one of the four "operating" sprinklers on the previous line.

If the branch lines have insufficient sprinklers to meet the required elongation, the design area should be extended to include sprinklers on additional branch lines supplied by the *same* cross main.

Now that the shape of the design area has been defined, we will consider the location of the design area.

DEAD-END SYSTEMS: The physically remote (from the riser) area of the system is normally the hydraulically remote area. In our example of the "1.2 times the square root of the area" rule, we casually stated that "the extra sprinkler would be placed…adjacent to any one of the four operating sprinklers." This has been stated, in different words, in a note in the annex of NFPA 13. While the location of the "extra" sprinkler or sprinklers is fairly academic in a grid, it makes a difference in a dead-end system. Consider the following examples:

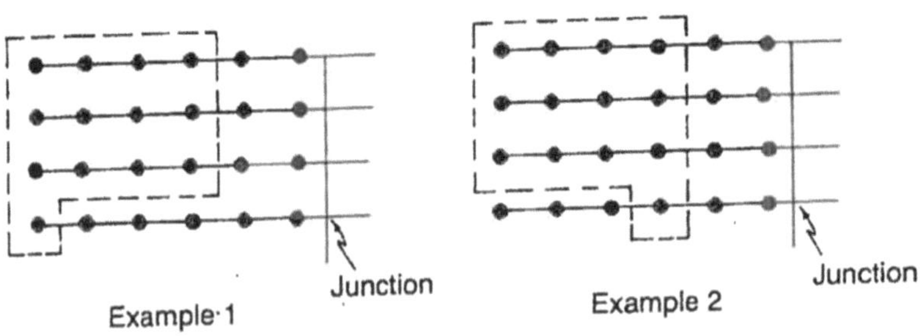

The thirteenth sprinkler could be placed at either of the locations shown above or anywhere in between.

Which of the above examples is more hydraulically demanding? At first glance, you might think it was Example 1, because the extra sprinkler is more remote physically. Actually, the most demanding location is Example 2. Why? Understand that the "extra" head must be balanced, in terms of pressure, at the junction of the branch line on which the extra head is located and the cross main. The governing pressure is that which has been calculated from the flow through the twelve sprinklers on the three downstream branch lines carried back to this junction. The balancing pressure from the branch line with the extra sprinkler is the sum of the pressure at the flowing sprinkler, which produces the associated flow, and the friction loss resulting from that associated flow carried back to the junction. Since the friction loss is maximized in Example 1, the pressure at the flowing sprinkler is minimized. Thus, the flow from the "extra" sprinkler is minimized and, therefore, the total design flow is minimized. On the other hand, the flow from the extra sprinkler in Example 2 is maximized, increasing the pressure as this flow is carried back through the system.

LOOPED SYSTEMS: As pointed out later on in the discussion of loops, the hydraulically remote point on a loop is halfway around the loop, assuming all of the pipes are of the same size. When the loop lacks symmetry, the equivalent pipe lengths of the elbows and tees must be included to find the halfway point. When there are different pipe sizes in the loop, you can convert all of the pipes to one equivalent pipe size to find the hydraulic halfway point.

GRIDDED SYSTEMS: The hydraulically most remote area of a grid is seldom obvious, because it is usually not the physically most remote area.

Things got off on the wrong foot here because, up through the 1976 Edition, NFPA 13 showed the physically remote area, the opposite comer of the grid from where the supply came in, as the hydraulically remote area in the two examples they provided. Most grid calculations of this era followed this, although it was incorrect.

Refer to the grid schematic on page 140. This is a typical grid. The supply is entering at the lower right. The physically remote area would be the upper left. Note the actual location of the hydraulically remote area. The computer program that calculated this system made successive calculations, starting with the area at the extreme left, then moving the area one sprinkler at a time to the right. As the area moved to the right, the friction loss between the grid entry point on the lower right and the operating area increased until the area was moved one head to the right of the position shown. In this manner, it was proven that the area shown is the hydraulically most remote, and this conforms to the verification requirement previously quoted. Although some people have tried to devise one, there is no simple "rule of thumb" that can be used to determine the location. This subject will be discussed further when we get to the grid.

By now, it should be clear that the selection of the hydraulically most remote area is not always a simple matter. As with everything else we talk about in this book, there may be times when the simple rules of NFPA 13 are not enough, when evaluation and good judgment are called for.

# Flow Velocity as a Constraint

A maximum flow rate of 16 feet per second in underground pipe and 32 feet per second in above-ground pipe is sometimes suggested as good practice. The origin of these numbers is obscure.

Some years ago, it was speculated that concern for scouring of cement-lined pipe was the source of the underground number. Another possible source was an ancient (and totally outdated) calculation by the American Water Works Association of the optimal flow rate, weighing pumping expenses versus required pipe sizes. As for the above-ground number, it has been speculated that somebody might have simply doubled the underground number! In truth, things like that sometimes happen. So much for the engineering profession.

At one time, these numbers appeared in NFPA 409, Aircraft Hangars, but somewhere along the way, they were quietly dropped. A 20 feet per second, rather than 32 feet per second, limit on above-ground piping is recommended in some quarters.

In 1980, the committee on NFPA 14, Standpipe and Hose Streams, proposed limiting the velocity in standpipe and supply piping to 16 feet per second, which was a bit curious since they were addressing above-ground piping. The proposal was subsequently withdrawn.

One insurance concern noted, on the pragmatic level, that with high velocities (near 32 feet per second), there will be a substantial reduction in density available at the most remote sprinkler if an additional sprinkler opens on the branch line beyond those calculated as opposed to a relatively small change in density with low velocities (less than 20 feet per second).

The equivalent pipe lengths for fittings that are in general use start to lose their validity at high flow rates, but how high is "high", we do not know.

Also, the Hazen–Williams formula yields less accurate results outside of the "normal" range of flows rates. Of course, the accuracy within the normal rate of flows is disputed in some quarters, as noted in the section on "Friction Loss of Water Flowing in a Pipe."



H. S. Wass Jr. and R. P. Fleming P.E., *Sprinkler Hydraulics*,
https://doi.org/10.1007/978-3-030-02595-3_18

NFPA 750, "Standard for the Installation of Water Mist Fire Protection Systems", first published in 1996, addresses the use of water in the form of a mist. The droplet size of the mist was appropriately defined, distinguishing it from the conventional sprinkler systems that are discussed in this book. They defined low-, intermediate-, and high-pressure systems. In low-pressure systems, "the distribution piping is exposed to pressures of 175 psi or less." Thus low-pressure systems were considered to be within the pressure range of NFPA 13 sprinkler systems and intermediate- and high-pressure systems out of the NFPA 13 allowable pressure range. They stated that hydraulic calculations for low-pressure systems with no additives could be calculated using the Hazen–Williams formula for friction loss and the velocity pressure formula. While consideration of velocity pressure is optional in NFPA 13, they seemed to suggest that it is mandatory. The Darcy–Weisbach method was set forth for intermediate- and high-pressure systems.

This subject was revisited by the NFPA 750 committee in 1999 and, Hazen–Williams was permitted for intermediate- and high-pressure systems when the flow velocity did not exceed 25 feet per second. Extensive comments on this subject were published, including data developed by Clyde Wood comparing Darcy–Weisbach and Hazen–Williams. (Clyde Wood of Automatic Sprinkler Company of America could be considered the father of sprinkler hydraulics in that his published material was the primary source on the subject prior to the introduction of hydraulic design into NFPA 13.) Clyde Wood's data suggest that the two formulas track closely at flow velocities between 5 and 18 feet per second with the accuracy of Hazen–Williams diminishing as you move away from that range.

The flow rate can be calculated by the following equation:

$$v = \frac{0.4085Q}{d^2} \tag{7}$$

where

$V$ is the flow velocity in feet per second; $Q$ is the flow rate in gallons per minute, and
$d$ is the internal pipe diameter in inches

NPFA 13 was silent on this subject until the 1999 Edition of NFPA 13. Apparently, a sprinkler contractor, when designing an addition to an existing system, came up with a flow in excess of 32 feet per second and the authority having jurisdiction told him to keep his flows at or below that rate. This led to a proposal to NFPA 13. The NFPA 13 committee agreed to the following additional appendix sentence:

It is not necessary to restrict the water velocity when determining friction loss using the Hazen–Williams formula.

The NFPA 13 committee tends to consider the flow velocity "self-regulating" because of the exponential increase in friction loss as the flow rate increases. This harmless affirmation of their previous silence will not necessarily change the requirements of some authorities having jurisdiction.

While the flow velocity is "self-regulating", whether or not it is sufficiently self-regulating is an open question.

The following table lists the maximum allowable flows, in gpm, for both 20 feet per second and 32 feet per second for the common sprinkler piping.

| Schedule 40 | | | Schedule 10 | | Schedule 5 | |
|---|---|---|---|---|---|---|
| Pipe size | 20 | 32 | 20 | 32 | 20 | 32 |
| 1″ | 54 | 86 | 59 | 94 | 69 | 110 |
| 1 ¼″ | 93 | 149 | 102 | 163 | 125 | 183 |
| 1 ½″ | 127 | 203 | 139 | 222 | 153 | 245 |
| 2″ | 209 | 335 | 228 | 365 | 247 | 395 |
| 2 ½″ | 299 | 478 | 340 | 544 | | |
| 3″ | 461 | 738 | 520 | 833 | | |
| 3 ½″ | 616 | 986 | 692 | 1,108 | | |
| 4″ | 794 | 1,270 | 889 | 1,422 | | |
| 5″ | 1,247 | 1,996 | 1,373 | 2,197 | | |
| 6″ | 1,801 | 2,882 | 1,979 | 3,166 | | |

While we are happy to publish this table, there are no inherent physical properties that make 20 and 32 critical numbers. But they may be useful reference points.

In the 2016 Edition of NFPA 13, stronger language against velocity limitations was provided as follows:

> 23.4.1.4 Unless required by other NFPA standards, the velocity of water flow shall not be limited when hydraulic calculations are performed using the Hazen-Williams or Darcy-Weisbach formulas.

# Calculating a Dead-End Sprinkler System

Having discussed all of the elements of calculating a sprinkler system, we will put this all together and make some actual calculations. For simplicity, we will ignore velocity pressure the first time around.

Refer to page 92 for a schematic of the system. Let us calculate a density of 0.20 gpm/sq. ft. over the hydraulically most remote 1800 sq. ft. The sprinkler spacing is 10 ft. × 10 ft. = 100 sq. ft. Therefore, the design area, 1800 sq. ft., must encompass 1800 ÷ 100 = 18 sprinklers. At the time this example was made up, NFPA 13 stated that the branch lines on both sides of the cross main should be included in the remote area, and the design area became the last two pairs of branch lines, which have a total of 18 sprinklers. While this requirement for the remote area changed long ago, the example still serves our purpose.

We will start our calculations at the end sprinkler of the most remote branch line, which has been labeled "1". Refer to pages 93 and 94 for the calculations. We have already determined that the sprinkler spacing, or area "covered" by each sprinkler, is 100 sq. ft. Multiplying 100 sq. ft. by the density of 0.20 gpm/sq. ft., we determined that the flow from the end sprinkler "1" should be 20 gpm. Now, it is necessary to determine the pressure required at this sprinkler to deliver this flow. Since $Q = k\sqrt{p}$, $\sqrt{p} = \frac{Q}{k}$, and $p = \left(\frac{Q}{k}\right)^2 = \left(\frac{20}{5.6}\right)^2 = 12.8\,\text{psi}$. Next, calculate the friction loss from "1" to "2". The friction loss through 1″ pipe, with 20 gpm flowing, is 0.130 psi/ft. Multiplying by the length, 10 ft., the friction loss is 1.3 psi, which is added to 12.8, the pressure at the end sprinkler, to get the pressure at sprinkler 2. Having calculated a pressure of 14.1 psi at sprinkler 2, we can calculate the flow from this sprinkler, $Q = 5.6\sqrt{14.1} = 21.0\,\text{gpm}$.

Adding the flow from "2" to the flow from "1", calculate the friction loss between "2" and "3".

You should be able to follow the calculations up to the point where a total of 85.9 gpm is shown flowing from the four sprinklers on this branch line. Now, we calculate the friction loss to a point at the top of the 2-in. riser nipple, taking into account the "T" at the top of the riser nipple. This involves 5 ft. of pipe and the



H. S. Wass Jr. and R. P. Fleming P.E., *Sprinkler Hydraulics*,
https://doi.org/10.1007/978-3-030-02595-3_19

equivalent pipe length for the tee. Note that we use the equivalent pipe length for a 1-½″ tee (8 ft.), not a 2-in. tee. Thus, we arrive at a flow of 85.9 gpm at 19.8 psi at the top of the riser nipple. We will return to these figures shortly.

At sprinkler 4, we had a total flow of 85.9 gpm at 16.7 psi. We can assume this flow and pressure at sprinkler 8, which is the equivalent sprinkler, the fourth sprinkler from the end, on the 5-sprinkler branch line. Then, calculate to the point at the top of the 2-in. riser nipple, as we did for the 4-sprinkler branch line. At this point, we have a flow of 110.4 gpm at 24.1 psi. Thus, 24.1 psi at the top of the riser nipple is needed to deliver the desired flow from the end sprinkler of the 5-sprinkler line. Therefore, it will be necessary to increase the flows from the sprinklers in the 4-sprinkler branch line to the point where, if recalculated, the pressure would also be 24.1 psi, rather than the 19.8 psi originally calculated. We can approximate the revised total flow from the 4-sprinkler branch line by using the equation:

$$\frac{Q_2}{Q_1} = \sqrt{\frac{P_2}{P_1}} \text{ or } Q_2 = Q_1\sqrt{\frac{P_2}{P_1}} \tag{8}$$

where $Q_1$ and $P_1$ are the flow and pressure values originally calculated and $Q_2$ and $P_2$ are the flow and pressure values we are adjusting to.

$P_2$ is 24.1 and we solve for $Q_2$, which turns out to be 94.8 gpm.

Combining the flows in the two branch lines yields a total flow of 205.2 gpm at 24.1 psi at the top of the riser nipple. Calculating the friction loss through the 1-in. riser nipple, the 90-degree elbow, and 10 ft. along the cross main to the next set of branch lines, we arrive at a pressure of 29.8 psi.

Note that we have ignored the elevation change through the 1-foot riser nipple. If all flowing sprinklers are at approximately the same level, and we are assuming that to be the case in this example, it is best to take the total elevation differences between the sprinklers assumed to be flowing and the base of the riser when you get to the base of the riser. Since we are going to calculate a branch line "k", it is more accurate to leave elevation out of it.

The simple method for calculating the flow from this second pair of branch lines is to treat these branch lines and their associated riser nipple as a single orifice and apply the equation for flow through an orifice:

$$Q = k\sqrt{p}$$

This approach is not entirely accurate (see discussion elsewhere of The Use and Abuse of the "k"), but it is good enough, the only practical approach for most hand calculations since the more rigorous alternative involves trial and error. This "k" approach need not be looked upon favorably if it is incorporated into a computer program, however. Perhaps we have a right to demand the extra effort of trial and error from a computer.

To apply the formula $Q = k\sqrt{p}$, we must know the "k". This pair of branch lines differs from the pair of branch lines previously calculated in that there is a tee at the base of the riser nipple, rather than a 90-degree elbow. Therefore, we will repeat the calculations through the riser nipple that we made for the end pair of branch lines using a tee rather than a 90-degree elbow. The resultant pressure is 28.0 psi and the flow remains the same, 205.2 gpm. These are the numbers we need to calculate the "k". The "k", of course, is equal to $Q \div \sqrt{p}$ and it turns out to be 38.779. We can now calculate the flow from the second set of branch lines, treating them as if they were one big sprinkler, using the calculated pressure of 29.8 psi at the base of this "sprinkler." The flow is 211.7 gpm which, combined with 205.2, gives a total flow through the design area of 416.9 gpm. From here, it is a simple matter to calculate friction loss back to the base of the riser and on out to the street, resulting in a required pressure of 44.9 psi.

On pages 95 and 96, the same calculations have been made with velocity pressure taken into account. Observe that the required flow and pressure are lower, demonstrating the fact that the error is on the conservative side when velocity pressure is ignored. You will note that pressure was carried to only one decimal place in the first example, whereas the pressure is carried to two decimal places when velocity pressure is included. It is common practice to carry the pressure to two decimal places. Although it implies a nonexistent precision, it has some merit in that occasionally the rounding-off errors will not balance out but accumulate.

On pages through is an example of hydraulic calculations with a pitched roof where the elevation difference between flowing sprinklers must be taken into account. Velocity pressure is included. Note that the elevation change between sprinkler 3 and the cross main is ignored in the branch-line calculations. This is in line with our previous recommendation that *all* elevation changes be entered into the calculations at the riser *except* when flowing sprinklers are involved. The elevation of 16.67 ft. added at the riser represents the elevation difference between sprinkler 3 and the base of the riser.

While it does not affect the validity of the calculation method we are illustrating, it is important to note that NFPA 13 prescribes a 30% increase in the design area where the pitch of the roof or ceiling exceeds 2 in. in 12 in. In other words, if the 0.25/2000 design in our example was the appropriate design in the past, the standard calls for 0.25/2600. This rule first appearing in the 1996 Edition of NFPA 13 makes a certain amount of sense since heat tends to move toward the highest level. Thus a fire starting toward the low end of the pitched roof would tend to open sprinklers away from the fire in the high area.

It should be noted, however, that beginning with the 2007 Edition of NFPA 13, a section was added to the storage protection rules that simply states that all of the storage protection criteria are intended to apply to buildings with ceiling slopes not exceeding 2 in 12 (16.7%) unless otherwise modified. This limitation was placed in the standard in recognition of the fact that the high-piled storage protection criteria were essentially developed based upon testing conducted under flat ceilings, and there is uncertainty over whether the 30% area increase would adequately address the effects of the ceiling slope.

The 30% increase in design area for sloped ceilings for other NFPA applications matches the 30% increase in design area traditionally required for dry pipe systems as compared to wet pipe systems. Because water delivery is delayed in the operation of dry pipe systems, it is expected that the fire can grow larger and subsequently open more sprinklers before fire control is established. For NFPA 13 applications involving a dry system under a sloped ceiling, the area increases must be compounded, making the protection of attics especially challenging.

Another notable NFPA 13 rule affecting the size of the design area is the minimum 3,000 sq. ft. design area where unsprinklered combustible concealed spaces are present in the building, although there are a number of special exceptions that apply.

Before leaving the subject of calculating branch lines, you may wish to refer to an easy way of making reasonably accurate calculations on conventional pipe schedule sprinkler systems, explained on pages 107 through 111.

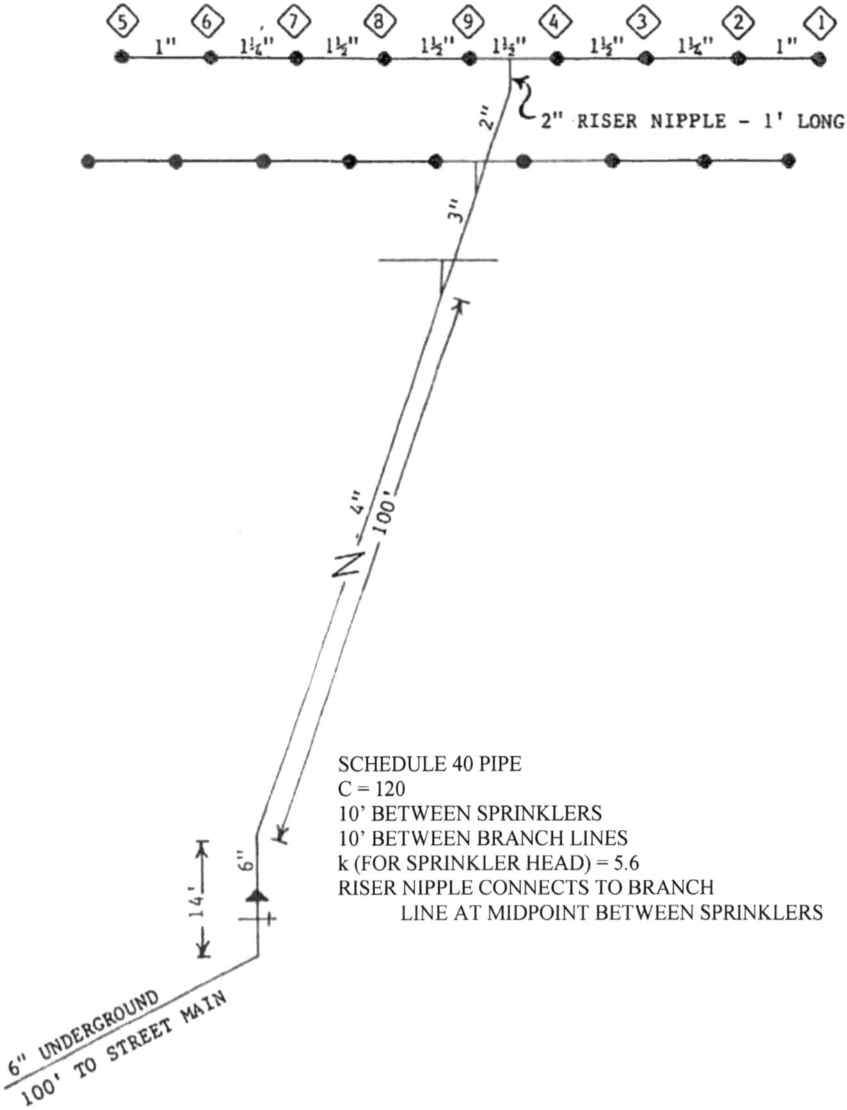

## HYDRAULIC CALCULATIONS

NAME DEAD END SYSTEM WITHOUT VELOCITY PRESSURE

ADDRESS

SPRINKLERS

AREA 1800 SQ. FT. - 18 HEADS (10 FEET BETWEEN ~~HEADS~~ ON BRANCH LINES, 10 FEET BETWEEN BRANCH LINES) 0.20 DENSITY

SHEET 1 OF 2
LOC. NO.
DATE
BY
Sched. 40 pipe C= 120 K= 5.6

| Nozzle Ident. & Location | Flow in G.P.M | Pipe Size | Pipe Ftgs. & Devices | Equiv. Pipe Length | Friction Loss PSI/Ft. | Pressure Summary | Normal Pressure | Notes |
|---|---|---|---|---|---|---|---|---|
| (1) | q.<br>Q. 20.0 | 1 | | L. 10<br>F.<br>T. 10 | .130 | Pt. 12.8<br>Pe. --<br>Pf. 1.3 | Pt.<br>Pv.<br>Pn. | $P = \left(\frac{Q}{K}\right)^2 = \left(\frac{20}{5.6}\right)^2 = 12.8$ |
| (2) | q. 21.0<br>Q. 41.0 | 1¼ | | L. 10<br>F. --<br>T. 10 | .129 | Pt. 14.1<br>Pe. --<br>Pf. 1.3 | Pt.<br>Pv.<br>Pn. | $Q = k\sqrt{P} = 5.6\sqrt{14.1} = 21.0$ |
| (3) | q. 22.0<br>Q. 63.0 | 1¼ | | L. 10<br>F. --<br>T. 10 | .135 | Pt. 15.4<br>Pe. --<br>Pf. 1.3 | Pt.<br>Pv.<br>Pn. | $Q = 5.6\sqrt{15.4} = 22.0$ |
| (4) | q. 22.9<br>Q. 85.9 | 1¼ | T | L. 5<br>F. 8<br>T. 13 | .239 | Pt. 16.7<br>Pe. --<br>Pf. 3.1 | Pt.<br>Pv.<br>Pn. | $Q = 5.6\sqrt{16.7} = 22.9$ |
| | q.<br>Q. | | | L.<br>F.<br>T. | | Pt. 19.8<br>Pe. --<br>Pf. -- | Pt.<br>Pv.<br>Pn. | |
| (8) | q.<br>Q. 85.9 | 1¼ | | L. 10<br>F. --<br>T. 10 | .239 | Pt. 16.7<br>Pe. --<br>Pf. 2.4 | Pt.<br>Pv.<br>Pn. | |
| (9) | q. 24.5<br>Q. 110.4 | 1¼ | T | L. 5<br>F. 8<br>T. 13 | .381 | Pt. 19.1<br>Pe. --<br>Pf. 5.0 | Pt.<br>Pv.<br>Pn. | $Q = 5.6\sqrt{19.1} = 24.5$<br>BALANCE FLOWS |
| | q. 94.8<br>Q. 205.2 | RN 2 | ELBOW | L. 1<br>F. 5<br>T. 6 | .355 | Pt. 24.1<br>Pe. --<br>Pf. 2.1 | Pt.<br>Pv.<br>Pn. | $\frac{Q_2}{Q_1} = \sqrt{\frac{P_2}{P_1}}$<br>$Q_2 = 85.9\sqrt{\frac{24.1}{19.8}} = 94.8$ |
| | q.<br>Q. | 2 | | L. 10<br>F. --<br>T. 10 | .355 | Pt. 26.2<br>Pe. —<br>Pf. 3.6 | Pt.<br>Pv.<br>Pn. | |
| | q.<br>Q. | | | L.<br>F.<br>T. | | Pt. 29.8<br>Pe. --<br>Pf. -- | Pt.<br>Pv.<br>Pn. | |
| Calc. "k" next set of BRANCH LINES | q.<br>Q. 205.2 | RN 2 | T | L. 1<br>F. 10<br>T. 11 | .356 | Pt. 24.1<br>Pe. --<br>Pf. 3.9 | Pt.<br>Pv.<br>Pn. | $k = \frac{205.2}{\sqrt{28.0}} = 38.779$ |
| | q. 211.7<br>Q. 416.9 | | | L.<br>F.<br>T. | | Pt. 28.0<br>Pe.<br>Pf. | Pt.<br>Pv.<br>Pn. | $Q = 38.779\sqrt{29.8} = 211.7$ |

# HYDRAULIC CALCULATIONS

NAME
ADDRESS
AREA

SHEET 2 OF 2
LOC. NO.
DATE
BY
C =
K =

| Nozzle Ident. & Location | Flow in G.P.M | Pipe Size | Pipe Ftgs. & Devices | Equiv. Pipe Length | Friction Loss PSI/Ft. | Pressure Summary | Normal Pressure | Notes |
|---|---|---|---|---|---|---|---|---|
| | q | 3" | | L 10 | .193 | Pt 29.8 | Pt | |
| | | | | F -- | | Pe -- | Pv | |
| | Q 416.9 | | | T 10 | | Pf 1.9 | Pn | |
| | q | 4" | | L 100 | .051 | Pt 31.7 | Pt | |
| | | | ELBOW | F 10 | | Pe -- | Pv | |
| | Q | | | T 110 | | Pf 5.6 | Pn | |
| | q | 6" | ACV=25 | L 14 | .007 | Pt 37.3 | Pt | |
| | | | E=14 | F 42 | | Pe 6.5 | Pv | $P_e$ = 15 x .433 = 6.5 |
| | Q | | GATE=3 | T 56 | | Pf .4 | Pn | |
| | q | 6" UG | | L 100 | .005 C=140 | Pt 44.2 | Pt | |
| | | | T | F 30 | | Pe -- | Pv | |
| | Q | | | T 130 | | Pf .7 | Pn | |
| | q | | | L | | Pt 44.9 | Pt | |
| | | | | F | | Pe | Pv | |
| | Q | | | T | | Pf | Pn | |
| | q | | | L | | Pt | Pt | |
| | | | | F | | Pe | Pv | |
| | Q | | | T | | Pf | Pn | |
| | q | | | L | | Pt | Pt | |
| | | | | F | | Pe | Pv | |
| | Q | | | T | | Pf | Pn | |
| | q | | | L | | Pt | Pt | |
| | | | | F | | Pe | Pv | |
| | Q | | | T | | Pf | Pn | |
| | q | | | L | | Pt | Pt | |
| | | | | F | | Pe | Pv | |
| | Q | | | T | | Pf | Pn | |
| | q | | | L | | Pt | Pt | |
| | | | | F | | Pe | Pv | |
| | Q | | | T | | Pf | Pn | |
| | q | | | L | | Pt | Pt | |
| | | | | F | | Pe | Pv | |
| | Q | | | T | | Pf | Pn | |
| | q | | | L | | Pt | Pt | |
| | | | | F | | Pe | Pv | |
| | Q | | | T | | Pf | Pn | |

## HYDRAULIC CALCULATIONS

NAME Preceding example taking velocity pressure into account

ADDRESS

AREA

NOTE: AQ means ASSUMED FLOW

SHEET 1 OF 2
LOC. NO.
DATE
BY
C = 120
K = 5.6

| Nozzle Ident. & Location | Flow in G.P.M | Pipe Size | Pipe Ftgs. & Devices | Equiv. Pipe Length | Friction Loss PSI/Ft. | Pressure Summary | Normal Pressure | Notes |
|---|---|---|---|---|---|---|---|---|
| <br>(1) | q<br>Q 20.0 | 1 | | L 10<br>F —<br>T 10 | .130 | Pt 12.76<br>Pe —<br>Pf 1.30 | Pt<br>Pv<br>Pn | $P = \left(\frac{20}{5.6}\right)^2 = 12.76$ |
| (2) | q 20.6<br>Q 40.6 | 1¼ | | L 10<br>F —<br>T 10 | .127 | Pt 14.06<br>Pe —<br>Pf 1.27 | Pt 14.06<br>Pv .52<br>Pn 13.54 | See Table V, Appendix A<br>$P_v = (AQ)^2K$<br>AQ = 40.8<br>$P_v = (40.8)^2 \times 3.10 \times 10^{-4} = 0.52$<br>$q = 5.6\sqrt{13.54} = 20.6$ |
| (3) | q 21.5<br>Q 62.1 | 1½ | | L 10<br>F —<br>T 10 | .131 | Pt 15.33<br>Pe —<br>Pf 1.31 | Pt 15.33<br>Pv .63<br>Pn 14.70 | AQ = 61.5<br>q = 21.5 |
| (4) | q 22.0<br>Q 84.1 | 1½ | T | L 5<br>F 8<br>T 13 | .230 | Pt 16.64<br>Pe —<br>Pf 2.99 | Pt 16.64<br>Pv 1.18<br>Pn 15.46 | AQ = 84<br>q = 22.0 |
| | q<br>Q | | | L<br>F<br>T | | Pt 19.63<br>Pe<br>Pf | Pt<br>Pv<br>Pn | |
| <br>(8) | q<br>Q 84.1 | 1½ | | L 10<br>F —<br>T 10 | .230 | Pt 16.64<br>Pe —<br>Pf 2.30 | Pt<br>Pv<br>Pn | |
| (9) | q 23.1<br>Q 107.2 | 1½ | T | L 5<br>F 8<br>T 13 | .361 | Pt 18.94<br>Pe —<br>Pf 4.67 | Pt 18.94<br>Pv 1.91<br>Pn 17.03 | AQ = 107<br>q = 23.1 |
| | q 92.3<br>Q 199.5 | RN 2 | ELBOW | L 1<br>F 5<br>T 6 | .337 | Pt 23.63<br>Pe —<br>Pf 2.02 | Pt<br>Pv<br>Pn | $Q_2 = 84.1\sqrt{\frac{23.63}{19.63}} = 92.3$ |
| | q<br>Q | 2 | | L 10<br>F —<br>T 10 | .337 | Pt 25.65<br>Pe —<br>Pf 3.37 | Pt<br>Pv<br>Pn | |
| | q<br>Q | | | L<br>F<br>T | | Pt 29.02<br>Pe<br>Pf | Pt<br>Pv<br>Pn | |
| | q<br>Q 199.5 | RN 2 | T | L 1<br>F 10<br>T 11 | .337 | Pt 23.63<br>Pe —<br>Pf 3.71 | Pt<br>Pv<br>Pn | $k = \frac{199.5}{\sqrt{27.34}} = 38.154$<br>Branch line "k" |
| | q 198.3<br>Q 397.8 | | | L<br>F<br>T | | Pt 27.34<br>Pe<br>Pf | Pt 29.02<br>Pv 2.01<br>Pn 27.01 | AQ = 398<br>$q = 38.154\sqrt{27.01} = 198.3$ |

# HYDRAULIC CALCULATIONS

NAME ______
ADDRESS ______
______
AREA ______
______

SHEET 2 OF 2
LOC. NO. ______
DATE ______
BY ______
C= ______
K= ______

| Nozzle Ident. & Location | Flow in G.P.M | Pipe Size | Pipe Ftgs. & Devices | Equiv. Pipe Length | Friction Loss PSI/Ft. | Pressure Summary | Normal Pressure | Notes |
|---|---|---|---|---|---|---|---|---|
| | q. | 3" | | L. 10 | .177 | Pt. 29.02 | Pt. | |
| | | | | F. — | | Pe. | Pv. | |
| | Q. 397.8 | | | T. 10 | | Pf. 1.77 | Pn. | |
| | q. | 4" | | L. 100 | .047 | Pt. 30.79 | Pt. | $P_e$ = 15 × .433 = 6.50 |
| | | | ELBOW | F. 10 | | Pe. | Pv. | |
| | Q. | | | T. 110 | | Pf. 5.17 | Pn. | |
| | q. | 6" | ACV 25 | L. 14 | .0064 | Pt. 35.96 | Pt. | |
| | | | OS & Y 3 | F. 42 | | Pe. 6.50 | Pv. | |
| | Q. | | ELBOW 14 | T. 56 | | Pf. .36 | Pn. | |
| | q. | 6" UG | | L. 100 | .0048 C=140 | Pt. 42.82 | Pt. | |
| | | | T | F. 30 | | Pe. — | Pv. | |
| | Q. | | | T. 130 | | Pf. .62 | Pn. | |
| | q. | | | L. | | Pt. 43.44 | Pt. | |
| | | | | F. | | Pe. | Pv. | |
| | Q. | | | T. | | Pf. | Pn. | |
| | q. | | | L. | | Pt. | Pt. | |
| | | | | F. | | Pe. | Pv. | |
| | Q. | | | T. | | Pf. | Pn. | |
| | q. | | | L. | | Pt. | Pt. | |
| | | | | F. | | Pe. | Pv. | |
| | Q. | | | T. | | Pf. | Pn. | |
| | q. | | | L. | | Pt. | Pt. | |
| | | | | F. | | Pe. | Pv. | |
| | Q. | | | T. | | Pf. | Pn. | |
| | q. | | | L. | | Pt. | Pt. | |
| | | | | F. | | Pe. | Pv. | |
| | Q. | | | T. | | Pf. | Pn. | |
| | q. | | | L. | | Pt. | Pt. | |
| | | | | F. | | Pe. | Pv. | |
| | Q. | | | T. | | Pf. | Pn. | |
| | q. | | | L. | | Pt. | Pt. | |
| | | | | F. | | Pe. | Pv. | |
| | Q. | | | T. | | Pf. | Pn. | |
| | q. | | | L. | | Pt. | Pt. | |
| | | | | F. | | Pe. | Pv. | |
| | Q. | | | T. | | Pf. | Pn. | |

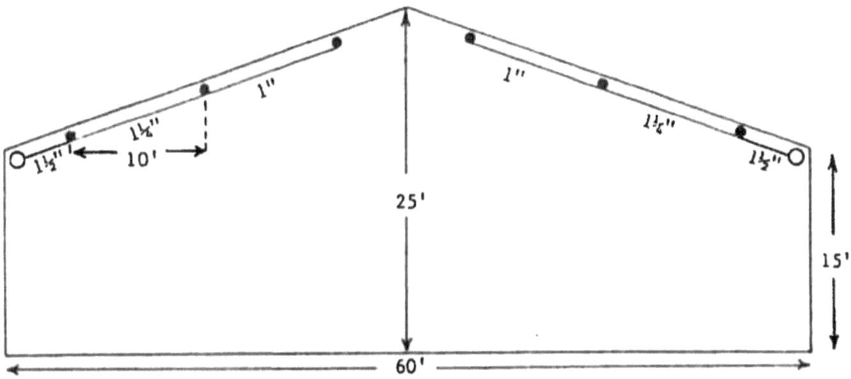


Design Criteria: 0.25 gpm/sq. ft. over 2000 sq. ft.
Sprinkler Spacing in Horizontal Plane: 10 ft.
Branch-Line Spacing: 10 ft.
Minimum Sprinkler Flow: 10 ft. × 10 ft. = 100 sq. ft. × 0.25 gpm/sq. ft. = 25 gpm
Elevation Change Between Sprinklers:

$$\frac{10}{30} = \frac{x}{10}, x = \frac{10}{3} = 3.333\ldots\text{ft.}$$

Distance Between Sprinklers:

$$\sqrt{10^2 + \left(\frac{10}{3}\right)^2} = 10.54\,\text{ft.}$$

Number of Flowing Sprinklers: $2000 \div 100 = 20$

Elongation of Operating Area along Branch Line: $1.2\sqrt{2000} = 53.7 \div 10 = 6$ sprinklers.

Since there are only 3 sprinklers on each branch line, extend the design area to additional branch lines supplied by the same cross main.

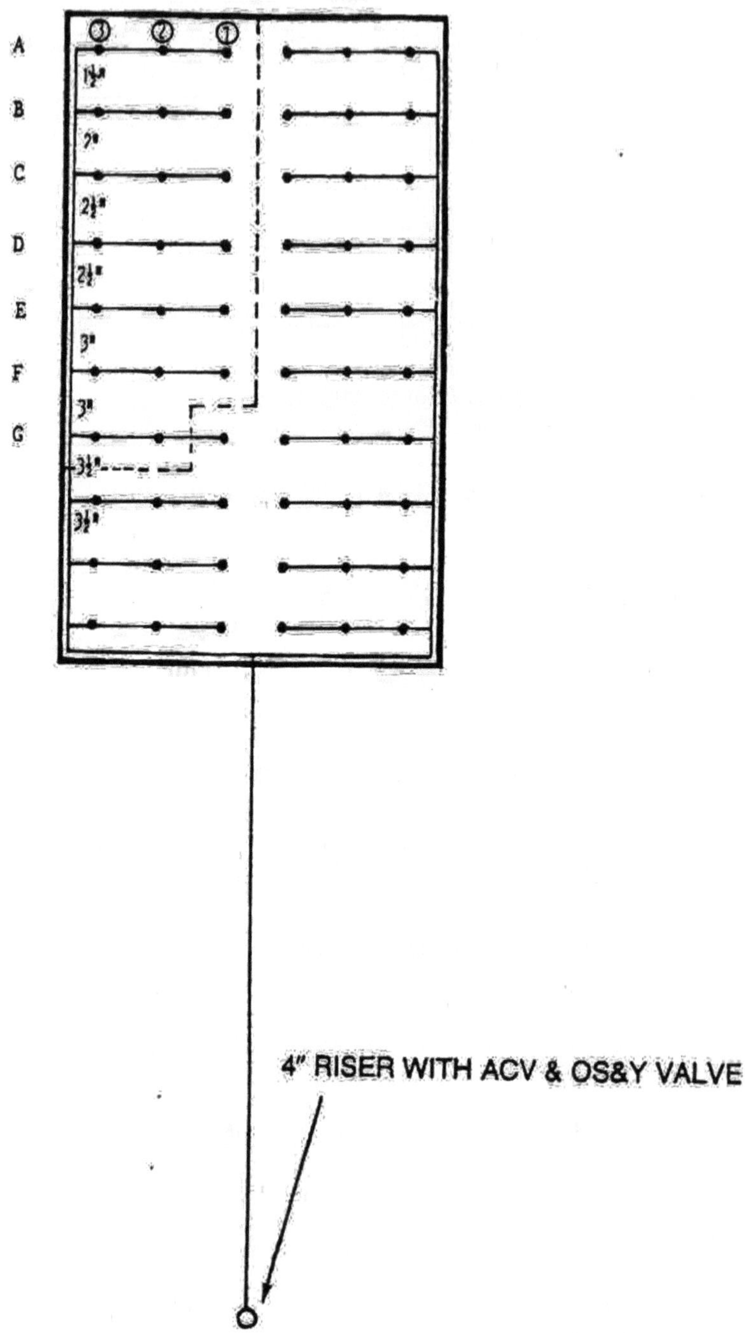
A
B
C
D
E
F
G
1½"
2"
2½"
2½"
3"
3"
3½"
3½"
4" RISER WITH ACV & OS&Y VALVE

## HYDRAULIC CALCULATIONS

NAME: CALCULATIONS BENEATH A PITCHED ROOF
ADDRESS:
AREA:

SHEET 1 OF 2
LOC. NO.
DATE
BY
C = 120
K = 5.6

| Nozzle Ident. & Location | Flow in G.P.M | Pipe Size | Pipe Ftgs. & Devices | Equiv. Pipe Length | Friction Loss PSI/Ft. | Pressure Summary | Normal Pressure | SCHEDULE 40 PIPE Notes |
|---|---|---|---|---|---|---|---|---|
| | | | | L. 10.54 | | Pt. 19.93 | Pt. | |
| 1 | q. 25.0 | 1" | | F. | .197 | Pe. 1.44 | Pv. | $P_e$ = .433x3.333 = 1.44 |
| | Q. 25.0 | | | T. 10.54 | | Pf. 2.08 | Pn. | |
| | | | | L. 10.54 | | Pt. 23.45 | Pt. 23.45 | ASSUMED FLOW, AQ = 51 |
| 2 | q. 26.6 | 1¼" | | F. | .198 | Pe. 1.44 | Pv. .81 | $P_V$ = K(AQ)[2], SEE TABLE V |
| | Q. 51.6 | | | T. 10.54 | | Pf. 2.09 | Pn. 22.64 | IN APPENDIX A |
| | | | | L. 4.5 | | Pt. 26.98 | Pt. 26.98 | |
| 3 | q. 28.5 | 1½" | T | F. 8 | .210 | Pe. | Pv. 1.07 | AQ = 80 |
| | Q. 80.1 | | | T. 12.5 | | Pf. 2.63 | Pn. 25.91 | |
| | | | | L. 10 | | Pt. 29.61 | Pt. | BRANCH LINE "k" = |
| | q. | 1½" | | F. | .210 | Pe. | Pv. | $\frac{80.1}{\sqrt{29.61}}$ = 14.72 |
| A–B | Q. 80.1 | | | T. 10 | | Pf. 2.10 | Pn. | |
| B | q. 80.8 | | | L. 10 | | Pt. 31.71 | Pt. 31.71 | |
| | | 2" | | F. | .226 | Pe. | Pv. 1.60 | AQ = 161.1 |
| | Q. 160.9 | | | T. 10 | | Pf. 2.26 | Pn. 30.11 | |
| C | q. 83.5 | | | L. 10 | | Pt. 33.97 | Pt. 33.97 | |
| | | 2½" | | F. | .207 | Pe. | Pv. 1.81 | AQ = 244.9 |
| C–D | Q. 244.4 | | | T. 10 | | Pf. 2.07 | Pn. 32.16 | |
| D | q. 84.3 | | | L. 10 | | Pt. 36.04 | Pt. 36.04 | |
| | | 2½" | | F. | .357 | Pe. | Pv. 3.26 | AQ = 328.5 |
| D–E | Q. 328.7 | | | T. 10 | | Pf. 3.57 | Pn. 32.78 | |
| E | q. 90.0 | | | L. 10 | | Pt. 39.61 | Pt. 39.61 | |
| | | 3" | | F. | .194 | Pe. | Pv. 2.24 | AQ = 419.7 |
| E–F | Q. 418.7 | | | T. 10 | | Pf. 1.94 | Pn. 37.37 | |
| F | q. 91.0 | | | L. 10 | | Pt. 41.55 | Pt. 41.55 | |
| | | 3" | | F. | .279 | Pe. | Pv. 3.31 | AQ = 510.7 |
| F–G | Q. 509.7 | | | T. 10 | | Pf. 2.79 | Pn 38.24 | |
| G | q. 69.3 | | | L. 20 | | Pt. 44.34 | Pt. 44.34 | AQ = 579 |
| | | 3½" | | F. | .174 | Pe. | Pv. 2.38 | See next page for |
| | Q. 579.0 | | | T. 20 | | Pf. 3.48 | Pn. 41.96 | calculation of G (2 heads) |
| | | | 2E | L. 144 | | Pt. 47.82 | Pt. | |
| | q. | 4" | 1T | F. 40 | .094 | Pe. | Pv. | |
| | Q. 579.0 | | | T. 184 | | Pf. 17.30 | Pn. | |
| | | | ACV | L. 15 | | Pt. 65.12 | Pt. | |
| | q. | 4" | OS&Y | F. 19 | .094 | Pe. 7.22 | Pv. | $P_e$ = .433x16.67 |
| | Q. 579.0 | | | T. 34 | | Pf. 3.20 | Pn. | |

# HYDRAULIC CALCULATIONS

NAME CALCULATIONS BENEATH A PITCHED ROOF

ADDRESS

AREA

SHEET 2 OF 2
LOC. NO.
DATE
BY
C =
K =

| Nozzle Ident. & Location | Flow in G.P.M | Pipe Size | Pipe Ftgs. & Devices | Equiv. Pipe Length | Friction Loss PSI/Ft. | Pressure Summary | Normal Pressure | Notes |
|---|---|---|---|---|---|---|---|---|
| | q.<br>Q. | | | L.<br>F.<br>T. | | Pt. 75.54<br>Pe.<br>Pf. | Pt.<br>Pv.<br>Pn. | Continued from Page 1 |
| | q.<br>Q. | | | L.<br>F.<br>T. | | Pt.<br>Pe.<br>Pf. | Pt.<br>Pv.<br>Pn. | |
| G2 | q. 34<br>Q. 34 | 1¼" | | L. 10.54<br>F.<br>T. 10.54 | .091 | Pt. 36.86<br>Pe. 1.44<br>Pf. .96 | Pt.<br>Pv.<br>Pn. | Assume a flow of 34 from Head G2 |
| G3 | q. 34.7<br>Q. 68.7 | 1½" | T | L. 4.5<br>F. 8<br>T. 12.5 | .158 | Pt. 39.26<br>Pe.<br>Pf. 1.98 | Pt. 39.26<br>Pv. .80<br>Pn. 38.46 | AQ = 69 |
| | q.<br>Q. | | | L.<br>F.<br>T. | | Pt. 41.24<br>Pe.<br>Pf. | Pt.<br>Pv.<br>Pn. | Adjust to $P_n$ at G<br>$Q_G = 68.7\sqrt{\frac{41.96}{41.24}} = 69.3$ |
| | q.<br>Q. | | | L.<br>F.<br>T. | | Pt.<br>Pe.<br>Pf. | Pt.<br>Pv.<br>Pn. | |
| | q.<br>Q. | | | L.<br>F.<br>T. | | Pt.<br>Pe.<br>Pf. | Pt.<br>Pv.<br>Pn. | |
| | q.<br>Q. | | | L.<br>F.<br>T. | | Pt.<br>Pe.<br>Pf. | Pt.<br>Pv.<br>Pn. | |
| | q.<br>Q. | | | L.<br>F.<br>T. | | Pt.<br>Pe.<br>Pf. | Pt.<br>Pv.<br>Pn. | |
| | q.<br>Q. | | | L.<br>F.<br>T. | | Pt.<br>Pe.<br>Pf. | Pt.<br>Pv.<br>Pn. | |
| | q.<br>Q. | | | L.<br>F.<br>T. | | Pt.<br>Pe.<br>Pf. | Pt.<br>Pv.<br>Pn. | |
| | q.<br>Q. | | | L.<br>F.<br>T. | | Pt.<br>Pe.<br>Pf. | Pt.<br>Pv.<br>Pn. | |

# Relating Hydraulic Calculations to the Water Supply

We have gone through the calculation of a tree system on the preceding pages. Assuming a required density of 0.20 over 1800 sq. ft. and given piping of specified sizes and lengths, it was determined on page 94 that a flow of 416.9 gpm at 44.9 psi (ignoring velocity pressure) was needed at the street to deliver the specified density over the specified area of application. This information is not meaningful unless it is related to the water supply. This is easily accomplished by a graphical analysis plotting pressure versus flow, with flow scaled to the 1.85 power (remember the relationship between flow and pressure in the Hazen–Williams formula). Refer to the graph on the next page.

A "Characteristic Curve for Design Area," a straight line that has been established by two points, is graphed. The first point is at 6.5 with zero flow. Why 6.5 psi? This is to account for the elevation of the sprinkler system with respect to the water supply. A total of 6.5 psi for elevation changes was added in the preceding calculations. The second point is the 416.9 gpm at 44.9 psi that was calculated.

The results of a flow test taken in the street are shown beneath the graph, and assume that the static and residual hydrant is located adjacent to the point where the sprinkler underground ties into the street main. The flow hydrant is a nearby hydrant. The two points for the "Flow Test" curve are the 50 psi static at zero flow and 845 gpm at 31 psi.



H. S. Wass Jr. and R. P. Fleming P.E., *Sprinkler Hydraulics*,
https://doi.org/10.1007/978-3-030-02595-3_20

The point at which the "Flow Test" curve crosses the "Characteristic Curve" is the real-life situation. By happy coincidence, the curves cross at the calculated design point, meaning that if all of the sprinklers are operating in the 1800-square-foot design area, we will, indeed, obtain a density of 0.20 gpm/sq. ft. What are we to make of it if our design point falls above or below the "Flow Curve"? The answer will be found on the ensuing pages as a part of a simplified method for calculating pipe schedule systems.

Note one thing. An allowance for hose streams should always be made with calculated systems. To keep it simple, hose streams are ignored at this time. They will be discussed later.

We have also ignored something else in the interest of simplicity, which we will discuss here.

NFPA 13 specifies that the required flow and pressure "at base of riser" be included in the "Hydraulic Design Information Sign" (at one time referred to as the "Nameplate") that should be provided for all calculated systems. They do not define the term "base of riser," and we are not aware that it appears anywhere else in the standard. We will fill in the void by defining it as the point of connection with the underground main. We will leave it up to you to define it in the rare case when there is no underground main. We will, however, hold to our simple definition when a system is supplied by an overhead main from a remote underground main. The "base of the riser" reference point is in common use. In the above example, the base of riser pressure is 44.2 (the flow is the same, of course), but the calculations were carried out to the effective point of the hydrant flow test in the street. If you are working from the base of riser numbers on the nameplate, there is a gap between the effective point of the hydrant flow test and the base of the riser, which must be closed.

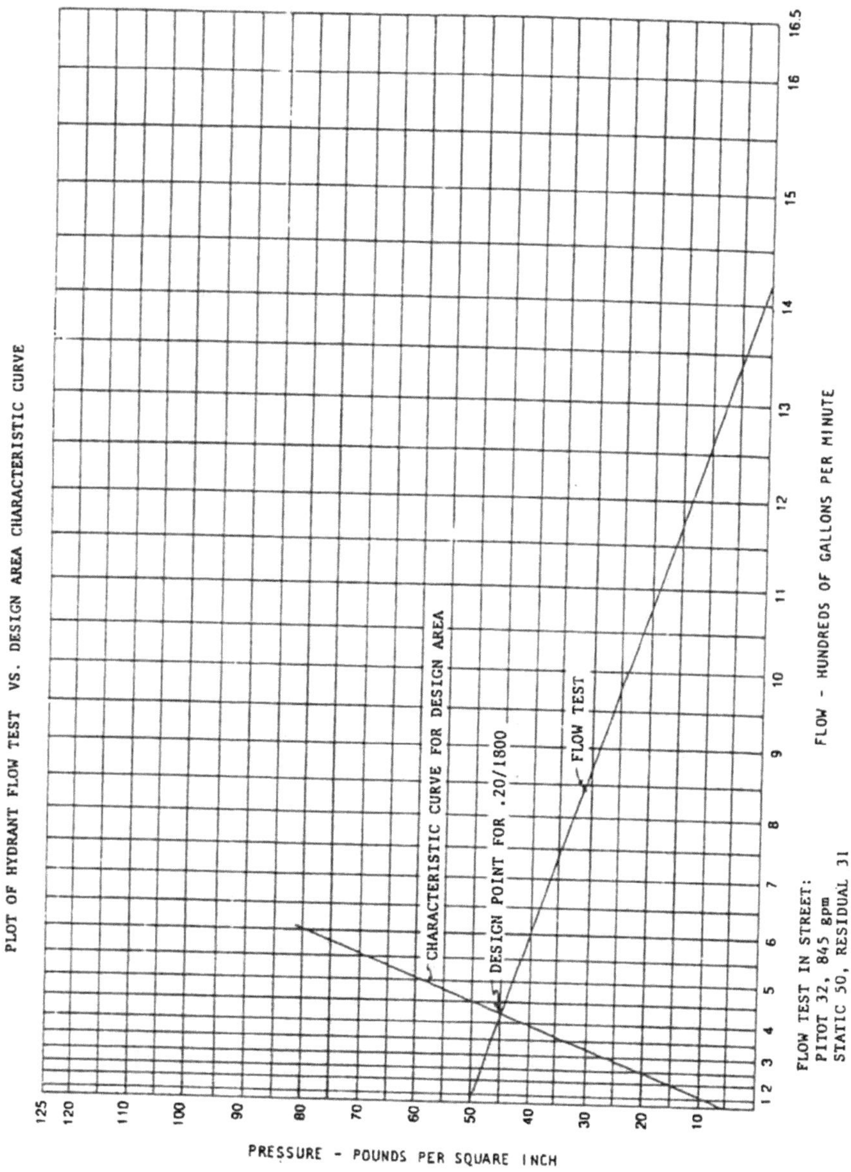
PLOT OF HYDRANT FLOW TEST VS. DESIGN AREA CHARACTERISTIC CURVE
CHARACTERISTIC CURVE FOR DESIGN AREA
DESIGN POINT FOR .20/1800
FLOW TEST
FLOW TEST IN STREET:
PITOT 32, 845 gpm
STATIC 50, RESIDUAL 31
FLOW - HUNDREDS OF GALLONS PER MINUTE
PRESSURE - POUNDS PER SQUARE INCH
125
120
110
100
90
80
70
60
50
40
30
20
10
1 2 3 4 5 6 7 8 9 10 11 12 13 14 15 16 16.5

The base of riser pressure can be adjusted by adding the friction loss for the design flow through the underground mains to the effective point of the flow test and adjusting for the elevation difference between the base of the riser and the point at which the static and residual pressure readings were taken. An alternative is to adjust the hydrant flow test to the base of the riser in a similar manner. It should be noted that the elevation difference between the base of the riser and the point where the static and residual pressures were measured is frequently overlooked, although often it is more significant than the friction loss between the base of the riser and the effective point of the flow test.

We have been using the phrase "effective point of the hydrant flow test" without explanation. Since occasionally there is confusion about this, four common examples appear on the next page.

Consider the difference between Examples 1 and 2. In Example 1, with a one-way feed, the residual pressure has meaning *only* at the point at which it is taken. The residual pressure at the junction of the sprinkler underground would be less, reflecting the friction loss in the street main between the junction and the point at which the residual pressure is taken.

In Example 2, the flow in the main is coming from both directions and the residual reading would be approximately the same at the junction of the sprinkler underground to the building. Therefore, we can consider the effective point to be at the junction.

In Example 3, there is no flow in the main to the left of the flow hydrant, so our residual reading at the building riser reflects the pressure in the street main at the point of connection of the flowing hydrant. In Example 4, water is flowing in the street main at the junction of the sprinkler underground, and the residual pressure at the riser reflects the pressure at that junction.

EXAMPLES OF THE EFFECTIVE POINT OF A HYDRANT FLOW TEST

1.

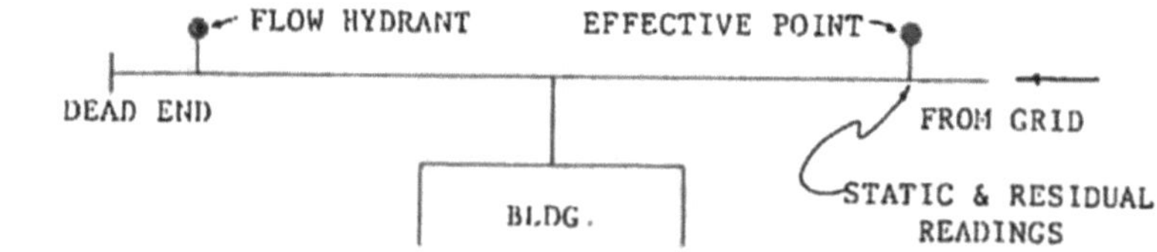


2.

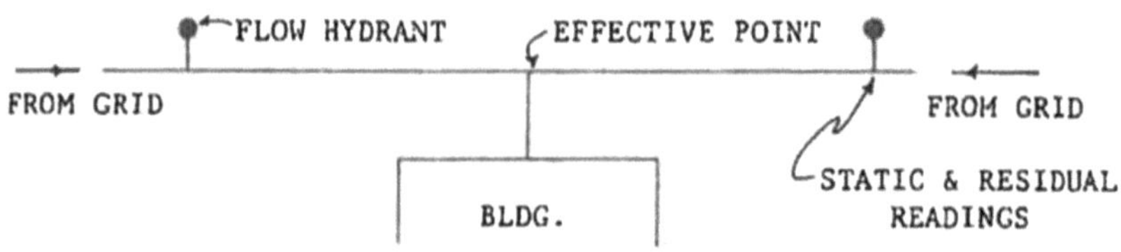

3.

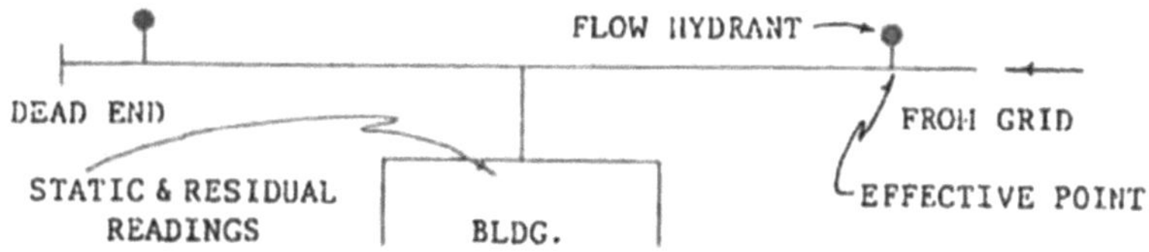


4.

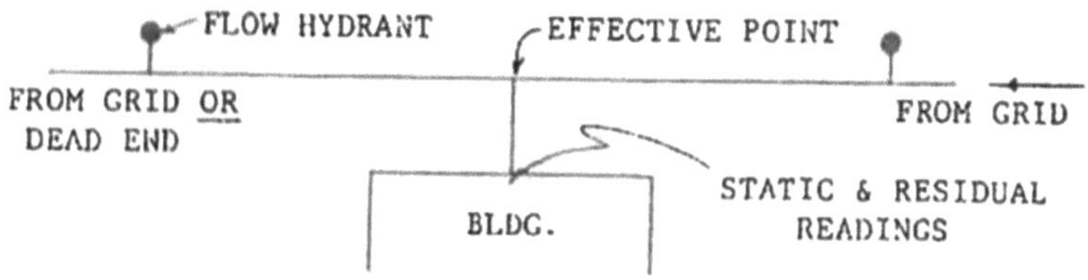


Note carefully that when matching up the water supply with the calculated sprinkler demand, the "gap" that must be closed for friction loss is between the base of the riser and the effective point of the hydrant flow test, whereas the "gap" that must be closed for elevation is between the base of the riser and the actual location of the gauge used for the static and residual pressure readings for the hydrant flow test.

A graphical solution is sufficiently accurate, but the junction point of the supply curve and the demand curve can be calculated with the following equation:

$$Q_J = \left[ \frac{P_S - P_E}{\frac{P_S - P_R}{Q_F^{1.85}} + \frac{P_D - P_E}{Q_D^{1.85}}} \right]^{0.54} \tag{9}$$

where

$Q_J =$ Flow at the junction of supply and design curves
$P_s =$ Static pressure
$P_R =$ Residual pressure on flow test
$P_D =$ Design pressure at the effective point of the flow test
$P_E =$ Height of sprinklers above the point at which the pressures were taken in the flow test, in psi
$Q_F =$ Flow, in flow test
$Q_D =$ Design flow

The associated pressure at the junction point, $P_J$, can be determined as follows:

$$P_J = P_S - \left( \frac{P_S - P_R}{Q_F^{1.85}} \right) Q_J^{1.85} \tag{10}$$

The use of the exponent 0.54 in the equation for $Q$ will produce a noticeable inaccuracy, and you may wish to use the repeating decimal 0.54054..., as previously discussed.

We will calculate $Q_1$ for the example that we have solved graphically:

$$Q_J = \left[\frac{50 - 6.5}{\frac{50-31}{845^{1.85}} + \frac{44.9-6.5}{416.8^{1.85}}}\right]^{0.54} = 414.1$$

and

$$P_J = 50 - \left(\frac{50 - 31}{845^{1.85}}\right)414.1^{1.85} = 44.9$$

While we are about it, here is the equation for calculating the flow at any desired pressure along the supply curve, with "$P$" representing the desired pressure

$$Q = Q_F - \left(\frac{P_S - P}{P_S - P_R}\right)^{0.54} \tag{11}$$

Refer to the flow test on the graph and calculate the available flow at 20 psi

$$Q = 845 - \left(\frac{50 - 20}{50 - 31}\right)^{0.54} = 1081$$

Conversely, the equation for calculating pressure, with $Q$ representing the chosen flow

$$P = P_S - \left(\frac{Q}{Q_F}\right)^{1.85} \times (P_S - P_R) \tag{12}$$

You may also wonder about calculating the junction point of the supply curve less a hose allowance and the demand curve. This is not as straightforward. If $Q$ is the flow at the junction point and $H$ is the hose stream allowance

$$\frac{P_P - P_E}{Q_D^{1.85}} \times Q^{1.85} + \frac{P_S - P_R}{Q_F^{1.85}} \times Q^{1.85} \times (Q+H)^{1.85} = P_S - P_E \tag{13}$$

$Q$ must be determined by trial and error. After determining $Q$, the associated pressure at this junction point can be calculated by:

$$P = P_E + \frac{P_D - P_E}{Q_D^{1.85}} \times Q^{1.85} \tag{14}$$

# A Simplified Method for Calculating Pipe Schedule Systems

Tables VII, VIII, IX, and X in Appendix A provide a simple method for calculating pipe schedule systems, either the current 2–3–5 schedule, the old 1–2–3 schedule, or an extra hazard pipe schedule. These tables are based upon a flow at the physically remote head (EHQ) of 25.0 gpm. The density that this end sprinkler flow translates to depends upon the sprinkler spacing. The top number in each pair of numbers in the table is the flow, in gpm, and the bottom number is the pressure, in psi.

For example, with a 2–3–5 pipe schedule and a 6-sprinkler branch line with 9-foot sprinkler spacing, the table provides a flow of 175.2 gpm at 40.5 psi. This is the pressure at the flowing sprinkler on the branch line closest to the cross main. Velocity pressure has been taken into account. While velocity pressure is included in the numbers in the tables, it can be *ignored* in subsequent branch-line calculations, both for simplicity and in partial compensation for the fact that, in many cases, not all of the sprinklers in the operating area are discharging as much as 25.0 gpm at the end sprinkler.

It should be noted that this table permits easy calculations where you do not wish to include all of the sprinklers on the branch line in the operating area. For example, if you wish to flow the 5 end sprinklers on an 8-sprinkler branch line, simply pick off the flow and pressure for a 5-sprinkler branch line and calculate the friction loss from that fifth sprinkler from the end back to the cross main, taking into account the intervening lengths of pipe and the fittings. You *cannot* use these tables when you have 9 or 10-sprinkler branch lines because the size of the second piece of piping from the end is different.



H. S. Wass Jr. and R. P. Fleming P.E., *Sprinkler Hydraulics*,
https://doi.org/10.1007/978-3-030-02595-3_21

We will illustrate the way you can put this table to practical use by an example. Refer to the ordinary hazard system depicted on page 110. Reasonably accurate calculations have been made for this system and about 709 gpm at 86 psi is required at the point indicated to deliver a minimum density of 0.30 gpm/sq. ft. to each sprinkler in the 2000 sq. ft. design area. The number above each discharging sprinkler shows the individual flows. You will note that the second from the end sprinkler on the second branch line is the one delivering the minimum required flow of 30 gpm, and the end sprinkler in the area is discharging 30.76 gpm. This is because velocity pressure is a significant factor with a relatively high flow through an ordinary hazard pipe schedule system.

Referring to Table VIII in Appendix A, the 2–3–5 table, select a flow of 144.8 gpm at 39.0 psi since there are 5 flowing sprinklers on a branch line and the distance between sprinklers is 11 ft. The total pressure at the first sprinkler in from the cross main on the branch line is 39.0 psi. We will now calculate the friction loss for the indicated flow back to the cross main.

There are 5.5 ft. of 1½″ pipe and a 1½″ tee, with EPL (equivalent pipe length) = 8 ft., for a total pipe length of 13.5 ft. Obtaining the friction loss per foot, 0.629 psi/ft, from the table in Appendix C and multiplying, we obtain a friction loss of 8.5 psi in the 1½″ pipe. There is 1 foot of 2-in. pipe in the riser nipple and an elbow, EPL = 5 ft., at the bottom for a total of 6 feet. 6 ft. × 0.187 psi/ft = 1.1 psi and the pressure at the cross main is 39.0 + 8.5 + 1.1 = 48.6 psi. The distance to the next branch line is 9.1 ft. 9.1 ft. × 0.187 psi/ft = 1.7, and the pressure at the base of the riser nipple for the second flowing branch line is 50.3 psi.

Now calculate a “k” for the second branch line. The end branch line had an elbow at the junction with the cross main whereas the other branch lines connect to the cross main with a tee. Substituting a tee, EPL = 10, for the elbow in the previous calculations, we get 11 ft. of 2-in., which multiplied by 0.187, yields 2.1 psi and a total pressure of 39.0 + 8.5 + 2.1 = 49.6 psi.

$$k = \frac{Q}{\sqrt{P}} = \frac{144}{\sqrt{49.6}} = 20.560$$

We can now calculate the flow from the second branch line.

We have already calculated that the pressure in the cross main at the connection point of the second branch line is 50.3 psi. (Remember that we ignore velocity pressure.) Therefore, the flow from the second branch line is $Q = K\sqrt{P} = 20.560\sqrt{50.3} = 145.8\,\text{gpm}$

Adding this to the flow from the first line, 144.8 + 145.8 = 290.6 gpm.

Friction loss to the third branch line: $0.285 \times 9.1 = 2.6\,\text{psi}. 2.6 + 50.3 = 52.9\,\text{psi}$. Therefore, $Q = 20.560\sqrt{52.9} = 149.5\,\text{gpm}$, and the total flow at this point is 290.6 + 149.5 = 440.1 gpm.

Friction loss to the fourth branch line: 0.213 × 9.1 = 1.9 psi 1.9 + 52.9 = 54.8 psi, and $Q = 20.560\sqrt{54.8} = 152.2\,\text{gpm}$, and the total flow is 440.1 + 152.2 = 592.3 gpm.

We now calculate the friction loss in 9.1 ft. of 3-in. pipe, which takes us to the end-point in the example: 0.369 × 9.1 = 3.4, and 3.4 + 54.8 = 58.2 psi.

What now? The flow from the table is based upon an end sprinkler flow of 25 gpm. All sprinklers are not flowing a minimum of 25 gpm, but we can ignore this without sacrificing very much accuracy. With a sprinkler spacing of 100 sq. ft., we have calculated for a density of 25 gpm/100 sq. ft. = 0.25 gpm/sq. ft. versus the 0.30 gpm/sq. ft. that we want. Since the extrapolation is modest, a graphical solution on 1.85 paper is sufficiently accurate, even though the 1.85 power is not an entirely valid relationship (see discussion of The Use and Abuse of the "k").

The calculated 592.3/58.2 point has been plotted on the graph on page 111 and the resultant characteristic curve is shown. Now divide the desired density by the density we have and multiply by the flow:

$$\frac{0.30}{0.25} \times 592.3 = 710.8\,\text{gpm}$$

Plotted on the characteristic curve, this flow seems to fall between 82 and 83 psi. Alternatively, we could calculate the pressure:

$$P_2 = P_1\left(\frac{Q_2}{Q_1}\right)^2 = 58.2\left(\frac{710.8}{592.3}\right)^2 = 83.8\,\text{psi}$$

Either way, the answer obtained by this simple method is fairly close to the reasonably "correct" answer of 708.9 gpm at 85.6 psi.

In common practice, when you calculate a pipe schedule system by this or any other method, while you are interested in a certain density over an area, such as 0.30 gpm per sq. ft. over 2000 sq. ft., you are really interested in calculating what density the system will actually deliver over the remote area and then you relate that to what you would like it to deliver.

What a system will actually deliver, of course, is related to the water supply. The graph on page 111 shows a water supply. The flow and pressure calculated in this example would normally be carried back to the public water supply but, for simplicity, assume that the water supply in the street has been adjusted to the end-point of this example (something that would never be done in practice). The flow at the point where the characteristic curve crosses the water supply, about 520 gpm, is what would actually discharge in the operating area. From this flow, we can approximate the density. It was previously determined that the flow at 592 gpm produced a density of 0.25 gpm per sq. ft. Multiplying 0.25 by the ratio of the flows: 520/592 × 0.25 = 0.22 gpm per sq. ft. (Average density buffs might note that the average density is 520/2000 = 0.26 vs. the 0.22 end-head density.)

We might wish to go one step further and determine what density could be obtained after making allowance for hose streams. For this, turn to page 171.

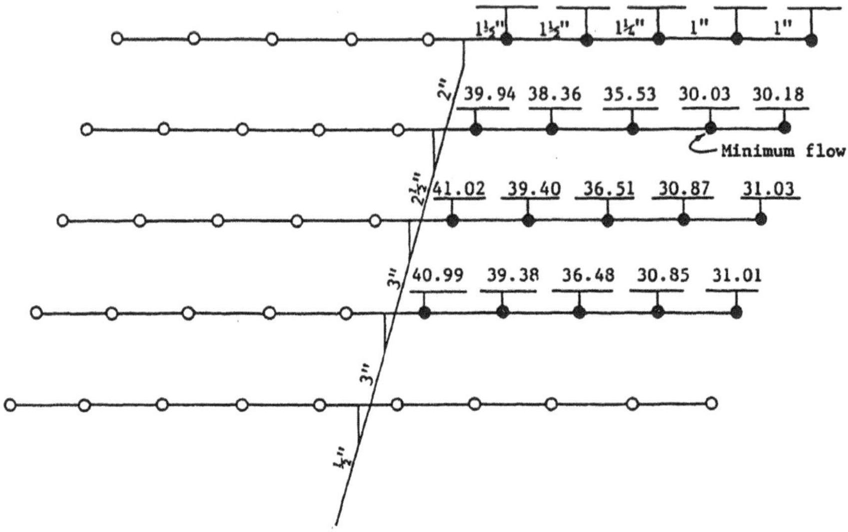


Sprinkler $k = 5.6$
Ordinary hazard pipe schedule, Schedule 40 pipe
Distance between sprinklers: 11 ft.
Distance between branch lines: 9.1 ft.
Area per sprinkler: 11 × 9.1 = 100.1 sq. ft.
Riser nipple, 1 ft. of 2 in. pipe connects to branch lines at the midpoint between sprinklers.

- ● Discharging sprinkler
- o Non-discharging sprinkler

*Wanted:* Density of 0.30 gpm per sq. ft. over 2000 sq. ft. Minimum sprinkler discharge: 0.30 gpm/sq. ft. × 100.1 sq. ft. = 30.03 gpm. The numbers above discharging sprinklers represent flows, in gpm.

RELATING THE DESIGN AREA CHARACTERISTIC CURVE TO THE WATER SUPPLY

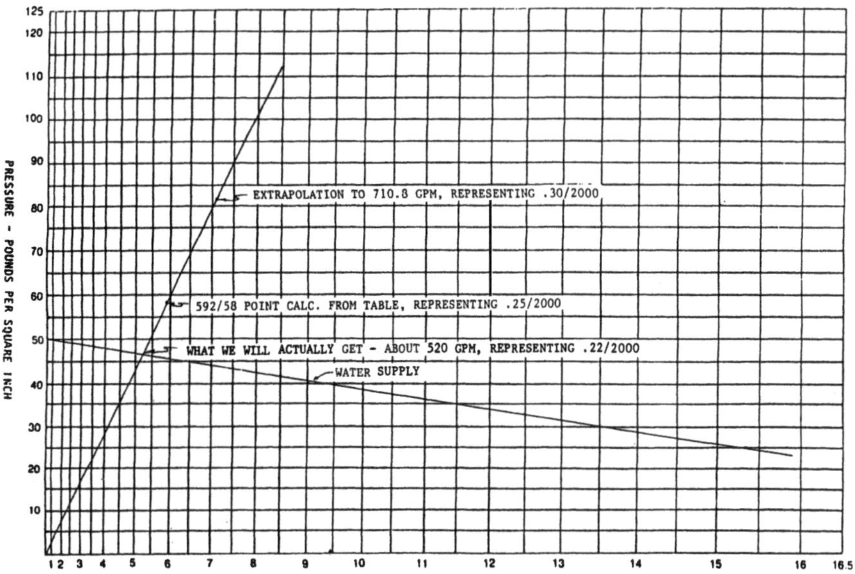

# The Loop

Rather than supplying water by the conventional "tree" system of dead-end lines, one simple modification that increases the efficiency of the system is the *loop*. The advantage of the loop derives from the fact that as flow decreases, the friction loss decrease is proportional to the flow to the 1.85 power. Consider a simple example:

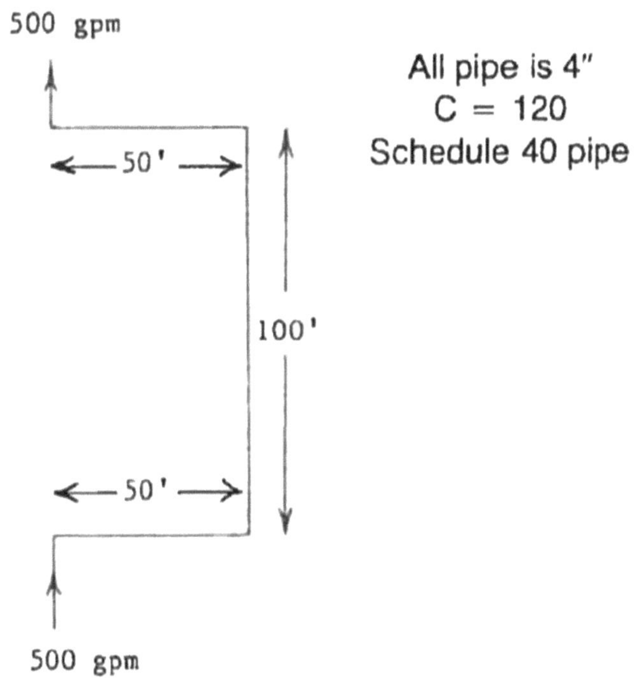


We are delivering 500 gpm to the area beyond. For simplicity, ignore the fittings at the entry and exit points and only consider the two elbows in between. The



H. S. Wass Jr. and R. P. Fleming P.E., *Sprinkler Hydraulics*,
https://doi.org/10.1007/978-3-030-02595-3_22

equivalent pipe length for these two elbows is 2 × 10 = 20, the actual pipe length is 50 + 100 + 50 = 200, and the total equivalent pipe length is 20 + 200 = 220 ft. The friction loss per foot for a 4-inch pipe with a flow of 500 gpm is 0.072 psi/ft. The loss between the entry and exit points is 220 ft. × 0.072 psi/ft = 15.8 psi.

Now consider what happens when we have a second, identical feed:

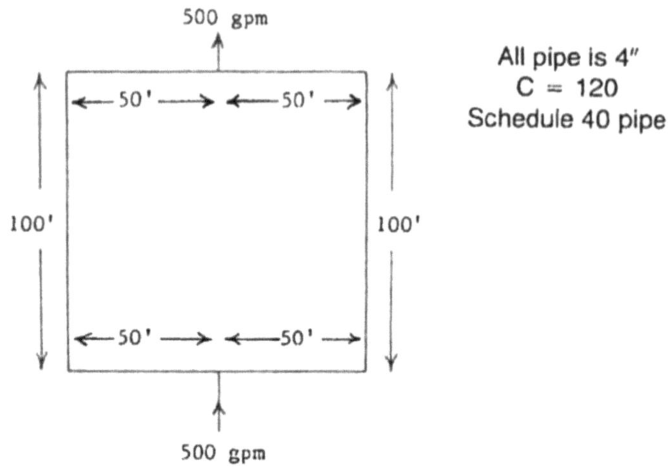


Because of the symmetry, it is obvious that the flow splits equally through each leg of the loop; that is, 250 gpm flows each way. The friction loss per foot for a 4-in. pipe with a flow of 250 gpm is 0.020 psi/ft. Multiplying 220 ft. × 0.020 psi/ft = 4.4 psi, compared to 15.8 psi with the single feed.

The preceding loop was simple because each leg was equal. What do we do when the two legs are not equal? Let us retain the same loop, but change the entry point:

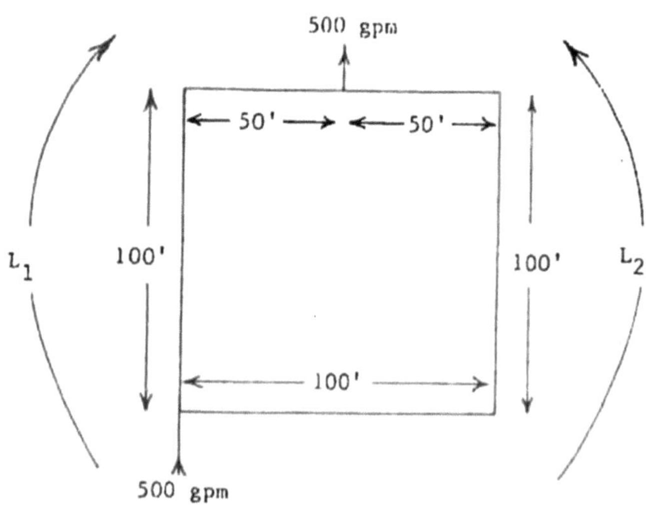

The friction loss through the two legs, $L_1$ and $L_2$, must be equal. We could assume flows and by trial and error gradually "zero in" on the answer. This can, however, be calculated directly.

In the section on friction loss, it was shown that the friction loss per foot is equal to a constant, which we will call "*K*," times the flow (*Q*) to the 1.85 power. Thus, the friction loss between two points can be expressed as $L_1KQ_1^{1.85} = L_2KQ_2^{1.85}$ where L is the length of the pipe.

Designate $L_1$ and $L_2$ as the two lengths of pipe and $Q_1$ and $Q_2$ and the respective flows. Since the friction loss through each leg is equal, $L_1KQ_1^{1.85} = L_2KQ_2^{1.85}$.

Since all of the pipes are 4-inch and of the same material, the *K*'s are equal and the equation becomes $L_1Q_1^{1.85} = L_2Q_2^{1.85}$.

Thus

$$\left(\frac{Q_1}{Q_2}\right)^{1.85} = \frac{L_2}{L_1}, \; or \; \frac{Q_2}{Q_1} = \left(\frac{L_2}{L_1}\right)^{0.54}$$

(The 0.54 power results from dividing the power of each side by 1.85.)

Since the total flow, *Q*, is the sum of $Q_1$ and $Q_2$, we can reduce this equation to one unknown, $Q_1$ by substituting $Q–Q_1$ for $Q_2$:

$$\frac{Q_1}{Q - Q_1} = \left(\frac{L_2}{L_1}\right)^{0.54}$$

Some algebraic manipulation produces the following:

$$Q_1 = \frac{Q}{\left(\frac{L_2}{L_1}\right)^{0.54} + 1} \tag{15}$$

We know that $Q = 500$

$L_1 = 100 + 50 + 10$ (elbow) $+ 20$ (tee) $= 180$

$L_2 = 20$ (tee) $+ 100 + 10$ (elbow) $+ 100 + 10$ (elbow) $+ 50 + 20$ (tee) $= 310$

Therefore,

$$Q_1 = \frac{500}{\left(\frac{180}{310}\right)^{0.54} + 1} = 286.4$$

and $Q_2 = 500 - 286.4 = 213.6$

To check this, we will now use the friction-loss tables:

$L_1$: 0.0256 psi/ft × 180 ft. = 4.61 psi

$L_2$: 0.0149 psi/ft × 310 ft. = 4.61 psi

Since the friction loss is identical through each leg, we have the correct division of flow.

Now that we know all about loops, we can enlarge on the general statements we made about the hydraulically remote area in the section on that subject. When a sprinkler system is supplied by a loop, the branch lines are commonly connected to a leg of the loop. Since a design area almost always includes multiple branch lines, the multiple outlets from the loop complicate things a bit. We will confine our discussion to the simplified case, the loops we have been working with where there is a single input to the loop and a single output.

The hydraulically remote outlet is the physically remote location on the loop when there is symmetry. The first loop example is symmetrical and the outlet, equidistant from the inlet along both legs, is at the hydraulically remote location.

The second loop example lacks symmetry. At the entry to the loop, one leg receives straight-through flow, whereas the other leg is connected with a tee. The most hydraulically remote outlet is again halfway around the loop, but "halfway" in terms of distance, which includes equivalent pipe length for fittings.

In the example of the second loop, the total equivalent lengths of the two legs were 180 ft. and 310 ft., respectively, a total of 490 ft. Each leg includes 20 ft. for the outlet tee, and, for the moment, we will deduct the outlet tees, leaving 450 ft. Half of 450 ft. is 225 ft. Follow Leg 2. The 20 ft. for the inlet tee, 100 ft. across the bottom, and 10 ft. for the elbow at the lower right add up to 130 ft., leaving 95 ft. to reach the midpoint of 225 ft. Thus, the hydraulically remote outlet location is 5 ft. below the upper right corner.

Put the outlet tee back in, and calculate the friction loss for the 245 ft. of equivalent pipe length in each leg. The total flow will be split equally through each leg.

From the table in Appendix C, the friction loss per foot for a flow of 250 gpm in 4-in. pipe is 0.020 psi/ft. Multiplying 0.020 psi/ft. $\times$ 245 ft. = 4.9 psi, somewhat more than when the outlet was at top-center.

In the foregoing examples, all of the pipes were the same size. If the loop contains more than one pipe size, the problem can be solved by the same method, but it is first necessary to convert all of the pipes to one equivalent pipe size. This involves selecting one pipe size to work with and converting all other pipe sizes to the equivalent length of the selected pipe size. This can be accomplished by calculating a factor to apply to the actual pipe length.

$$\text{FACTOR} = \left(\frac{D_2}{D_1}\right)^{4.87} \tag{16}$$

where

$D_1$ is the actual internal diameter
$D_2$ is the internal diameter of the pipe being converted to

If you have Schedule 40 or Schedule 10 pipe you can refer to Table VI of Appendix A rather than calculating it with Eq. 2.

Before we leave this section, we will develop another equation related to the loop illustrated on the bottom of page 121. All of the pipe in the loop is the same size. What would be the length of a single pipe of the same diameter that would carry the combined flows of the two legs of the loop and produce the same friction loss?

Referring to our derivation of Eq. 1 on page 122, it is apparent that $L_1Q_1^{1.85} = L_2Q_2^{1.85} = L_E(Q_1 + Q_2)^{1.85}$ and $\frac{Q_2}{Q_1} = \left(\frac{L_1}{L_2}\right)^{0.54}$ *or* $Q_2 = \left(\frac{L_1}{L_2}\right)^{0.54} \times Q_1$

Therefore

$$L_1Q_1^{1.85} = L_E\left[Q_1 + \left(\frac{L_1}{L_2}\right)^{0.54} \times Q_1\right]^{1.85}$$

$$L_E = \frac{L_1Q_1^{1.85}}{\left[Q_1 + \left(\frac{L_1}{L_2}\right)^{0.54} \times Q_1\right]^{1.85}}$$

or

$$L_E = \left[\frac{Q_1}{\left(\frac{L_1}{L_2}\right)^{0.54} \times Q_1 + Q_1}\right]^{1.85} \times L_1$$

and canceling out the *Q*'s

$$\text{Equivalent Length} = L_E = \left[\frac{1}{\left(\frac{L_1}{L_2}\right)^{0.54} + 1}\right]^{1.85} \times L_1 \qquad (17)$$

# Introducing…The Grid

We are ready to consider the grid. The favorable hydraulic characteristics of a gridded sprinkler system permit relatively small pipe sizes and uniform pipe sizing, which translate into lower costs. Therefore, grids predominate except where the building arrangement does not lend itself to reasonably regular sprinkler and branch line spacings, typically in highly compartmented buildings, such as hospitals.

A grid is really just a series of loops. In fact, it is really not a grid. In its infancy, descriptive words such as "ladder" or "railroad track" were used. "Grid" is short and simple, and it is the accepted term today, even if it is misleading to the outsider. A schematic of a typical grid can be found on page 140.

The normal grid consists of branch lines of uniform size throughout their length, and the pipe size is generally quite small (1-in. to 2-in. diameter). The branch lines are connected on each end to larger mains, one of which is connected to the water supply. There is no standard terminology. Terms such as "near" or "near side" main and "far" or "far side" main may be used, with the near side main being the one connected to the water supply. The main connected to the water supply may be descriptively called the "supply-side main" and the other main referred to as the "tie-in-side main." Whatever it is called, the main on the supply side is usually slightly larger than the main on the other side because the typical flows are larger.

Referring again to the schematic of a typical grid, you can see that with a good water supply it is capable of delivering a density of 0.4 gpm per sq. ft. over 2000 sq. ft., when the pressure is about 65 psi at the entrance to the grid despite having 1-1/2 in. branch lines more than 140 ft. long. When the remote area is discharging, all of the non-discharging branch lines are carrying water from the supply-side main to the tie-in-side main.

The grid is a very efficient means of supplying water to the sprinkler heads, because, unlike in a tree system, all of the piping in the system carries water to the flowing sprinklers. Everything else being equal, however, the reliability of the protection from a gridded system should be considered less than from a non-gridded system. While a slight elongation along the branch lines of the assumed operating area is currently required (see discussion of the hydraulically most remote area), it is



H. S. Wass Jr. and R. P. Fleming P.E., *Sprinkler Hydraulics*,
https://doi.org/10.1007/978-3-030-02595-3_23

possible for more than the contemplated number of sprinklers to open along a branch line. This could be caused by an unusual draft condition, an unanticipated flammable liquids fire, or the arrangement of a highly combustible commodity. As more sprinklers open along a small diameter branch line, there would be a fairly rapid decay of density.

Although the grid is very efficient, it is only efficient when there is water in the pipes, as in a wet pipe system. The problem discovered with the gridded dry system is the time it takes for water to move through the entire network of pipes while expelling air. In a gridded tree system, it is only necessary for the water to move from the dry pipe valve along a single path to the first sprinkler that opens. But in a grid, when just a single sprinkler is open, water flows through all of the pipes in the system. When the first water reaches the open sprinkler, the system will not yet be "up to speed"; that is, the total flow in the system will not have been established.

The first, if inadequate, acknowledgment of the problem surfaced in the 1980 Edition of NFPA 13 when the committee limited pipe volume to 500 gallons for gridded systems unless water could be delivered to the inspector's test pipe in not more than one minute. Finally, in the 1987 Edition of NFPA 13, gridded dry pipe systems were prohibited. No doubt, however, there are some old dry gridded systems still in existence. Although some water delivery time calculation programs have become commercially available in the intervening years, the dry grid is still prohibited by NFPA 13.

There is a variation of the standard grid where the side mains do not connect the ends of the branch lines, but are moved in a few sprinklers, resulting in short closed-end branch lines on each side of the grid and a reduction of the length of the branch lines between the side mains.

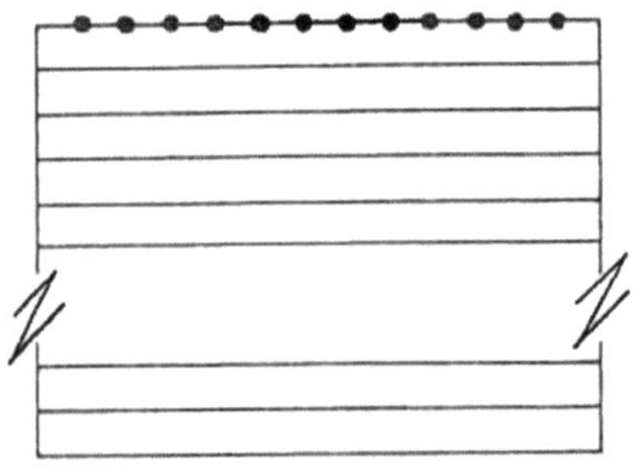

Conventional Grid

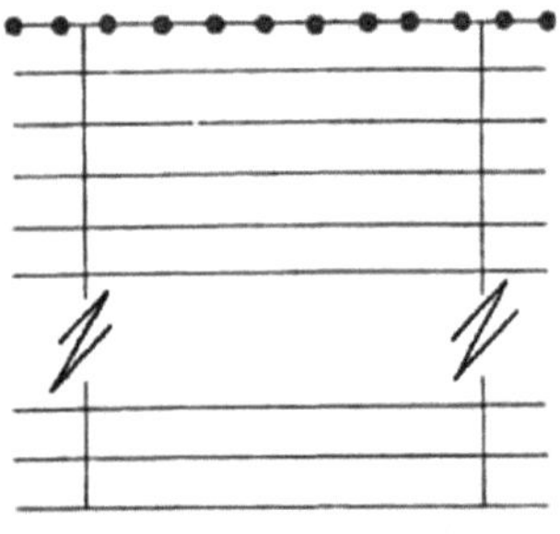

Grid with Outriggers

The branch lines extending outside of the "grid" are usually called outriggers, and the sprinklers are sometimes referred to as outboard sprinklers. The net result is slightly better hydraulics with no increase in pipe length or size and only a slight increase in the cost of fittings. Not all computer programs can easily analyze a design such as this. In most layouts, it is assumed that the remote area remains within the grid, and the short closed-end lines are simply ignored in the calculations. Obviously, however, as the side mains are moved closer together, at some point the hydraulically most remote area will jump over to include the closed-end branch lines. The only specific advice I can offer is to be alert to this possibility.

The "typical," "standard," or "simple" grid fits nicely into flat-roofed rectangular buildings with open areas. The following information gives a complete physical description of the "simple" grid:

1. Distance between sprinklers along the branch line.
2. Number of sprinklers on a branch line.
3. Distance, including the equivalent pipe length of fittings, from the near- or supply-side main to the first sprinkler on the branch line.
4. Distance, including the equivalent pipe length of fittings, from the far- or tie-in-side main to the first head on the branch line.
5. Distance between branch lines.
6. Number of branch lines.
7. Diameter of branch lines.
8. Diameter of near- or supply-side main.
9. Diameter of far- or tie-in-side main.
10. The location of the connection of the supply to the supply-side main.
11. The sprinkler "k."

In the real world, however, with irregular buildings, partitions, and obstructions, variations in Items 1–5 are common. Item 11 will vary in cases where sprinklers are on drop nipples or riser nipples. Once in a while a contractor will choose to vary Items 8 or 9 to meet the hydraulic requirements of the system. Also, one pertinent piece of information is *assumed* in the above listing: that all sprinklers are at the same elevation.

The following additional information permits calculations to be made:

1. "C" (120 for a wet system, 100 for a dry system).
2. Design density.
3. Design area.
4. Shape of the design area (elongation of 1.2 times the square root of the area according to NFPA 13, but some authorities specify 1.4 times the square root of the area).

One further assumption has been made: namely, that the design area should lie in the hydraulically remote area of the grid. This would not be true, for example, if the design criteria were for storage in an area of the grid that was not hydraulically remote.

Before getting into the heart of the grid, we will look at a few peripheral matters, the first uncommon, the second very common.

A special kind of loop problem occasionally arises in grids. Because of partitions, perhaps, an otherwise conventional grid might have something like this:

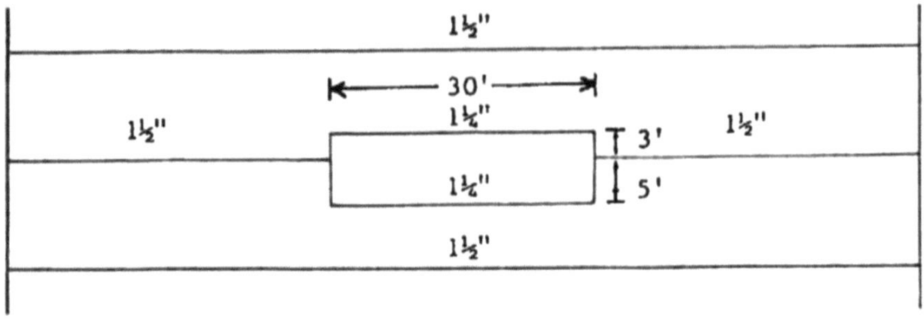


First Case

or this:

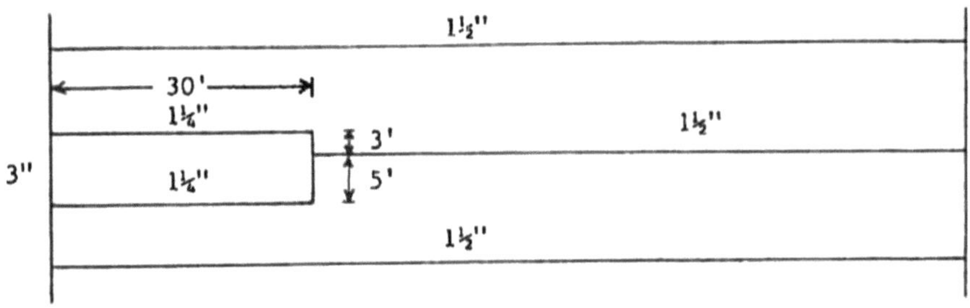


Second Case

Since the input to most computer grid programs will be much simpler, you may wish to convert the parallel 1-1/2 in. lines to an equivalent length of a single 1-1/2 in. line and tell the computer that the only irregularity is that this branch line has a different length. This is usually an acceptable approximation assuming there are no flowing sprinklers in these lines. Here is how to do it:

**First Case**:
Length of upper 1-1/4 in. line, including fittings:

$$6(\text{tee}) + 3 + 3(\text{elbow}) + 30 + 3(\text{elbow}) + 3 + 6(\text{tee}) = 54$$

Length of lower 1-1/4 in. line, including fittings:

$$6(\text{tee}) + 5 + 3(\text{elbow}) + 30 + 3(\text{elbow}) + 5 + 6(\text{tee}) = 58$$

$$\text{Equation 3 from page 117} \quad L_E = \left[\frac{1}{\left(\frac{L_1}{L_2}\right)^{0.54} + 1}\right]^{1.85} \times L_1$$

Let $L_1 = 54$ and $L_2 = 58$

It makes no difference which lengths you define as $L_1$ and $L_2$ although a slight difference may show up in the second decimal place unless you expand the approximate exponent, 0.54–0.54054, as discussed elsewhere.

$$L_E = \left[\frac{1}{\left(\frac{54}{58}\right)^{0.54} + 1}\right]^{1.85} \times 54 = 15.52\,\text{ft. of } 1\text{-}\frac{1}{4}\,\text{in. pipe}$$

Now convert the 1-1/4 in. pipe to 1-1/2 in. pipe (assume Schedule 40 pipe and see Table VI, Appendix A, to obtain the conversion factor).

$$15.52 \times 2.12 = 32.90\,\text{ft. of } 1\text{-}1/2\,\text{in. pipe}$$

**Second Case**:

Length of upper 1-1/4 in. line, including fittings:

$$6(\text{tee}) + 3 + 3(\text{elbow}) + 30 + 6(\text{tee}) = 48$$

Length of lower 1-1/4 in. line, including fittings:

$$6(\text{tee}) + 5 + 3(\text{elbow}) + 30 + 6(\text{tee}) = 50$$

The friction loss in the 3-in. line resulting from the assumption that the 1-1/2 in. line continues straight across is sufficiently small that it can be ignored.

Let $L_1 = 48$ and $L_2 = 50$

$$L_E = \left[\frac{1}{\left(\frac{48}{50}\right)^{0.54} + 1}\right]^{1.85} \times 48 = 13.59\,\text{ft. of } 1\text{-}1/4\,\text{in. pipe}$$

$$13.59 \times 2.12 = 28.8\,\text{ft. of } 1\text{-}1/2\,\text{in. pipe}$$

In both cases, the equivalent length of 1-1/2 in. pipe calculated above would be added to the length of actual 1-1/2 in. pipe and the equivalent length of the actual 1-1/2 in. fittings (2 1-1/2 in. tees in the first case and a single 1-1/2 in. tee in the second case).

Now let's look at a more common variation encountered in a grid. As previously stated, the supply is connected to one of the side mains. It may be connected at the end of the main.

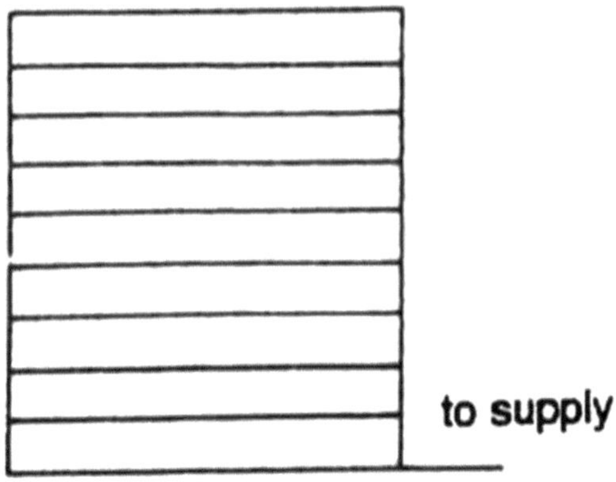


Many times the point of connection is somewhere else. Consider this example:

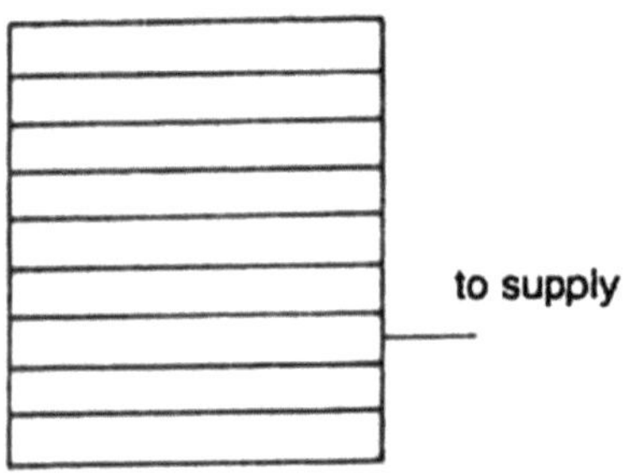


Schedule 10 pipe
Length of branch lines: 80 ft. Distance between branch lines: 10 ft. Size of branch lines: 1 ¼"
1 ½" riser nipples, 1 ft. long, connecting branch lines to side mains, tee at bottom, elbow at top
Size of supply-side main:
3" Size of far-side main: 2 1/2"
Supply connects to side main at midpoint between adjoining branch lines.

The hydraulically remote area is clearly somewhere along the branch lines at the top of the grid since the supply is connected to the side main below the midpoint of the side main. If the design area includes the top three branch lines, it would be apparent that the direction of flow in each segment of pipe would be as follows:

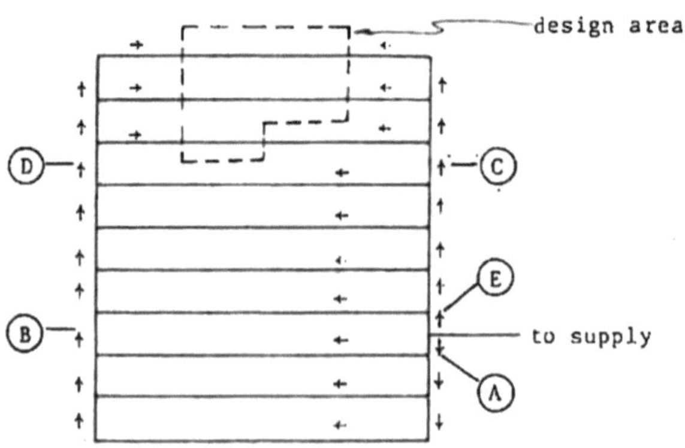

If you look at this piping schematic carefully, it can be seen that the sum of the flows in pipe segments C and D is equal to the total flow from the supply. It should also be obvious that the flow in pipe segment A is equal to the flow in pipe segment B.

We will now focus our attention on the portion of the grid below the point where the supply enters the side main. Since all of the water is entering this portion of the grid through one segment of pipe (A), and leaving through one segment of pipe (B), this portion of the grid can be reduced to a single equivalent pipe. The portion of the grid above the supply connection cannot be reduced in this fashion because there are two entry segments (B and E) and two exit segments (C and D).

Equation 17 can be used to determine the equivalent single pipe length of a loop.

$$L_E = \left[\frac{1}{\left(\frac{L_1}{L_2}\right)^{0.54} + 1}\right]^{1.85} \times L_1$$

Now we will apply this equation to the portion of the grid below the

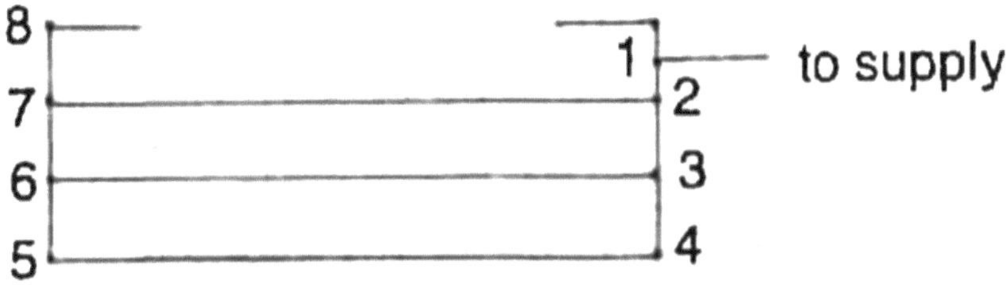


First we will consider the bottom loop, 3–4–5–6–3. The loop equation assumes all pipe of the same internal diameter, so we will convert the side mains to their equivalent lengths of 1-1/4 in. pipe, the branch line diameter. Referring to Table VI in Appendix A for Schedule 10 pipe,

For segment 3–4: 0.0188 × 10 = 0.19 ft.
For segment 5–6: 0.0531 × 10 = 0.53 ft.

Now calculate the total equivalent length of the branch lines (4–5 and 6–3).

80 ft. actual length plus 2 × (1 ft. actual length of riser nipple plus 6 ft. equivalent pipe length for the tee at the bottom plus 3 ft. equivalent pipe length for the standard elbow at the top) = 100 ft.

The loop consists of a lower leg, 3–4–5–6, and an upper leg, 3–6. We will let 3–4–5–6 be $L_1$ and 3–6 be $L_2$. $L_1 = 0.19 + 0.53 + 100 = 100.72$ ft. $L_2 = 100$ ft.

$$L_E = \left[\frac{1}{\left(\frac{L_1}{L_2}\right)^{0.54} + 1}\right]^{1.85} \times L_1 = \left[\frac{1}{\left(\frac{100.72}{100}\right)^{0.54} + 1}\right]^{1.85} \times 100.72 = 27.84\,\text{ft.}$$

Applying Eq. 17.

$$L_E = \left[\frac{1}{\left(\frac{L_1}{L_2}\right)^{0.54} + 1}\right]^{1.85} \times L_1 = \left[\frac{1}{\left(\frac{28.56}{100}\right)^{0.54} + 1}\right]^{1.85} \times 28.56 = 13.35\,\text{ft.}$$

Thus the loop 3–4–5–6–3 can be considered a single piece of 1-1/4 in. pipe 27.84 ft. long, and the portion of the grid below the supply now looks like this:

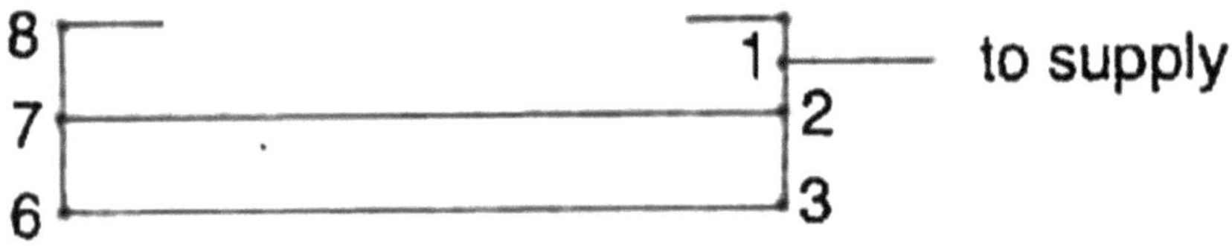


The loop consists of a lower leg, 2–3–6–7, and an upper leg, 2–7. Calling 2–3–6–7 $L_1$, $L_1 = 0.19 + 27.84 + 0.53 = 28.56$ and, again, $L_2 = 100$.

The lower portion of the grid has been reduced to this:

where 2–7 can be considered to be a branch line 13.35 ft. long. This simple branch line can now be used, in place of the three branch lines below the supply, in the actual grid calculations.

# The Grid… Getting to Know You

It is time to take a look at what happens in a grid. Perhaps the easiest way to do this is to examine what we might call a mini-grid. You will never encounter it in the real world, but the basic principles are the same in all grids. We will assume that the design area is two sprinklers.

CROSS SECTION

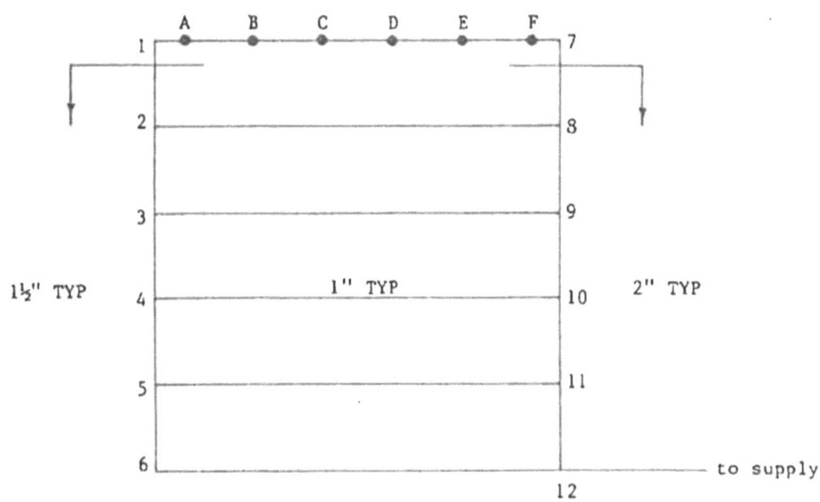


PLAN VIEW

Schedule 40 pipe



H. S. Wass Jr. and R. P. Fleming P.E., *Sprinkler Hydraulics*,
https://doi.org/10.1007/978-3-030-02595-3_24

c = 120
k = 5.6
Sprinkler spacing: 10 ft.
Branch line spacing: 12 ft.
Minimum discharge from each sprinkler: 25 gpm

We will approach this problem with complete ignorance beyond the basics of hydraulics covered in the earlier sections of the book. Since the supply enters the grid at the bottom, it is clear that the hydraulically remote "area," consisting of the two discharging sprinklers, is located on the top branch line, but we cannot be sure which two of the six sprinklers on that line are the most remote. Since the supply is connected to the right side, we will first look at the two sprinklers on the left side, A and B.

Table 4 of Appendix A provides the following friction loss constants:

| | |
|---|---|
| 1″ pipe | $5.099 \times 10^{-4}$ |
| 1-1/2" pipe | $6.330 \times 10^{-5}$ |
| 2" pipe | $1.875 \times 10^{-5}$ |

The total equivalent branch line length = 50 (distance from Sprinkler A to Sprinkler F) + 2 [4 + 2 + 5 (tee) + 2 (elbow)] = 50 + (2 × 13) = 76 equivalent feet of l-in. pipe.

The friction loss in each branch line =

$$76 \times 5.099 \times 10^{-4} Q^{1.85} = 0.03875\, Q^{1.85} \text{psi.}$$

The friction loss for each far main segment =

$$12 \times 6.33 \times 10^{-5} = 0.0007596\, Q^{1.85} \text{psi.}$$

The friction loss for each near main segment =

$$12 \times 1.875 \times 10^{-5} = 0.000225\, Q^{1.85} \text{psi.}$$

The next question is how the flow splits. Actually, this is the only question. The flow split governs everything that follows. We will make the simple assumption that all of the discharge from Sprinkler A flows from the left and all of the discharge from Sprinkler B flows from the right.

Pressure required at Sprinkler A to discharge 25 gpm:

$$P = \left(\frac{Q}{k}\right)^2 = \left(\frac{25}{5.6}\right)^2 = 19.93 \text{ psi.}$$

Calculate the friction loss from Sprinkler A to Node 2 (it is fairly common practice to refer to junction points in grids as nodes—sounds good).

$$\begin{aligned} &\text{A to 1}: 13 \times .0005099 \times 25^{1.85} &&= 2.56 \\ &\text{1 to 2}: 0.0007596 \times 25^{1.85} &&= \underline{0.29} \\ & &&2.85\ \text{psi} \end{aligned}$$

$$\text{Pressure at Node 2} = 19.93 + 2.85 = 22.78\ \text{psi}$$

Calculate the friction loss from Sprinkler B to Node 8
Pressure required at Sprinkler 8 (same as Sprinkler A): 19.93

$$\begin{aligned} &\text{B to 7}: 53 \times .0005099 \times 25^{1.85} &&= 10.42 \\ &\text{7 to 8}: 0.000225 \times 25^{1.85} &&= \underline{0.09} \\ &\text{Pressure at Node 8} &&= 30.44\ \text{psi} \end{aligned}$$

The flow in branch lines 2–8 must be such that the friction loss will be 30.44 − 22.78 = 7.66 psi.

We have already established that $P = 0.03875Q^{1.85}$ for each branch line. Therefore,

$$Q_{8-2} = \left(\frac{P}{0.03875}\right)^{0.54} = \left(\frac{7.66}{0.03875}\right)^{0.54} = 17.37\ \text{gpm}$$

Ignorant though we are, we do not believe that two-thirds of the flow from the left side would flow through the first transfer branch line, leaving only one-third of the flow for the remaining four branch lines. So we will start over again, *increasing* the flow to the left. Assume that 5 gpm from Sprinkler 8 flows from the left, leaving 20 gpm flowing from the right.

| | |
|---|---|
| Pressure required at Sprinkler B : | 19.93 |
| Friction loss from SprinklerB to Sprinkler A : 0.010 (From App. E, Friction Loss Table) × 10 | = 0.10 |
| Pressure at Sprinkler A : | 20.03 psi |
| Flow from Sprinkler A : | |
| $Q = k\sqrt{p} = 5.6\sqrt{20.03}$ | 25.06 |
| Add flow from Sprinkler B | 5.00 |
| | 30.06 gpm |

Calculate the friction loss from Sprinkler A to Node 2.

| | |
|---|---|
| Pressure required at Sprinkler A : | 20.03 |
| A to 1 : $13 \times 0.0005099 \times 30.06^{1.85}$ | 3.60 |
| 1 to 2 : $0.0007596 \times 30.06^{1.85}$ | 0.41 |
| Pressure at Node 2 | 24.04 psi |

Calculate the friction loss from Sprinkler B to Node 8 (flow is 20 gpm)

| | |
|---|---|
| Pressure required at Sprinkler B : | 19.93 |
| B to 7 : $53 \times 0.0005099 \times 20^{1.85}$ | 6.90 |
| 7 to 8 : $0.000225 \times 20^{1.85}$ | 0.06 |
| Pressure at Node 8 | 26.89 psi |

$$Q_{8-2} = \left(\frac{26.89 - 24.04}{.03875}\right)^{0.54} = 10.18\,\text{gpm}$$

This still looks a little high. Let's try 7 gpm flowing from the left to Sprinkler B.

| | |
|---|---|
| Pressure required at Sprinkler B : | 19.93 |
| Friction loss from SprinklerB to Sprinkler A : | |
| $0.019 \times 10 =$ | 0.19 |
| Pressure at Sprinkler A : | 20.21 psi |
| Flow from Sprinkler A : $Q = 5.6\sqrt{20.21}$ | 25.12 |
| Add flow from Sprinkler B | 7.00 |
| | 32.12 gpm |

| | |
|---|---|
| Pressure at Sprinkler A : | 20.12 |
| A to 1 : $13 \times .0005099 \times 32.12^{1.85}$ | 4.06 |
| 1 to 2 : $0.0007596 \times 32.12^{1.85}$ | 0.47 |
| Pressure at Node 2 | 24.65 psi |

| | |
|---|---|
| Pressure required at Sprinkler B : | 19.93 |
| B to 7 : $53 \times .0005099 \times 18^{1.85}$ | 5.68 |
| 7 to 8 : $0.000225 \times 18^{1.85}$ | 0.05 |
| Pressure at Node 8 | 25.66 psi |

$$Q_{8-2} = \left(\frac{25.66 - 24.65}{0.03875}\right)^{0.54} = 5.82\,\text{gpm}$$

This looks more reasonable, so we will carry it through to the supply

$$Q_{2-3} = 32.12 - 5.82 = 26.30\,\text{gpm}$$

$$P_{2-3} = 0.0007596 \times 26.30^{1.85} = 0.32$$
$$\text{Pressure at Node 3} = 24.65 + 0.32 = 24.97\,\text{psi}$$

$$Q_{8-9} = 18 + 5.82 = 23.82$$

$$P_{8-9} = 0.000225 \times 23.82^{1.85} = 0.08$$
$$\text{Pressure at Node 9} = 25.66 + 0.08 = 25.74\,\text{psi}$$

$$Q_{3-9} = \left(\frac{25.74 - 24.97}{0.03875}\right)^{0.54} = 5.02\text{gpm}$$

$$Q_{3-4} = 26.30 - 5.02 = 21.28\text{ gpm}$$

$$P_{3-4} = 0.0007596 \times 21.28^{1.85} = 0.22$$

Pressure at Node 4 $= 24.97 + 0.22 = 25.19$ psi

$$Q_{9-10} = 23.82 + 5.02 = 28.84\text{ gpm}$$

$$P_{9-10} = 0.000225 \times 28.84^{1.85} = 0.11$$

Pressure at Node 10 $= 25.74 + 0.11 = 25.85$ psi

$$Q_{4-10} = \left(\frac{25.85 - 25.19}{0.03875}\right)^{0.54} = 4.62\text{ gpm}$$

$$Q_{4-5} = 21.28 - 4.62 = 16.66\text{ gpm}$$

$$P_{4-5} = 0.0007596 \times 16.66^{1.85} = 0.14$$

Pressure at Node 5 $= 25.19 + 0.14 = 25.33$ psi

$$Q_{10-11} = 28.84 + 4.62 = 33.46$$

$$P_{10-11} = 0.000225 \times 33.46^{1.85} = 0.15$$

Pressure at Node 11 $= 25.85 + 0.15 = 26.00$ psi

$$Q_{5-11} = \left(\frac{26.00 - 25.33}{.03875}\right)^{0.54} = 4.66\text{ gpm}$$

$$Q_{5-6} = 16.66 - 4.66 = 12.00$$

$$P_{5-6} = 0.0007596 \times 12.00^{1.85} = 0.08$$

Pressure at Node 6 $= 25.33 + 0.08 = 25.41$ psi

$$Q_{11-12} = 33.46 + 4.66 = 38.12\text{ gpm}$$

$P_{11-12}$ The tee at the entrance of the supply to the grid should be considered in this leg. 10 ft. equivalent pipe length for the tee plus 12 ft. actual length = 22 ft.

$1.875 \times 10^{-5} \times 22 = 0.0004125$

$$P_{11-12} = 0.0004125 \times 38.12^{1.85} = 0.35$$
$$\text{Pressure at Node } 12 = 26.00 + 0.35 = 26.35\,\text{psi}$$

$$Q_{6-12} = Q_{5-6} = 12.00\,\text{gpm}$$

$$P_{6-12} = 0.03875 \times 12.00^{1.85} = 3.84$$
$$\text{Pressure at Node } 12 = 25.41 + 3.84 = 29.25\,\text{psi}$$

The pressure at Node 12 coming from the left side is 2.90 psi higher than the pressure coming from the right side. Therefore, we need to reduce the flow from the left to the discharging sprinklers. We will reduce the flow from the left to Sprinkler B from 7.00 to 6.70 gpm.

| | |
|---|---|
| Pressure required at Sprinkler B : | 19.93 |
| Friction loss from SprinklerB to Sprinkler A : | |
| $0.017 \times 10 =$ | 0.17 |
| Pressure at Sprinkler A : | 20.21 psi |
| Flow from Sprinkler A : $Q = 5.6\sqrt{20.10}$ | 25.11 |
| Add flow from Sprinkler B | 6.70 |
| | 31.81 gpm |

| | |
|---|---|
| Pressure at Sprinkler A : | 20.10 |
| A to 1 : $13 \times 0.0005099 \times 32.12^{1.85}$ | 3.99 |
| 1 to 2 : $0.0007596 \times 32.12^{1.85}$ | 0.46 |
| Pressure at Node 2 | 24.55 psi |

| | |
|---|---|
| Pressure required at Sprinkler B: | 19.93 |
| B to 7 : $53 \times 0.0005099 \times 18^{1.85}$ | 5.85 |
| 7 to 8 : $0.000225 \times 18^{1.85}$ | 0.05 |
| Pressure at Node 8 | 25.83 psi |

$$Q_{2-8}C\left(\frac{25.83 - 24.55}{0.03875}\right)^{0.54} = 6.61\,\text{gpm}$$

Since we have already gone through the step-by-step procedure we will skip a number of the intermediate steps (you can do them if you are interested) and pick up toward the end.

$$Q_{11-12} = 36.76 + 6.12 = 42.88\,\text{gpm}$$

$$P_{11-12} = 0.0004125 \times 42.88^{1.85} = 0.43$$

$$\text{Pressure at Node } 12 = 26.23 + 0.43$$
$$= 26.66\,\text{psi}\; Q_{6-12} = Q_{5-6} = 7.23\,\text{gpm}$$

$$P_{6-12} = 0.03875 \times 7.23^{1.85} = 1.51$$

$$\text{Pressure at Node 12} = 25.15 + 1.51 = 26.66\,\text{psi}$$

Success!! The pressure at Node 12 is 26.66 psi calculated from both directions.

With brute force and a lot of luck, we solved a simple grid quite easily. Obviously, some kind of method can be devised for a computer to make successive guesses as to where the flow splits based upon the previous trials. The goal is rapid convergence on the "right" split.

The traditional method for attacking the problem of calculating the flows in a network of the pipe is the "Hardy Cross Method." Much is heard about "Hardy Cross," but little is written with respect to sprinkler protection. In fact, some people wonder if "Hardy Cross" refers to one man or two, Hardy and Cross. "Hardy Cross" refers to Professor Hardy Cross, an American Civil Engineer (1885–1959), who used his method to analyze flows in gridded city water mains (true grids).

Take another look at the Hazen-Williams equation (see Eq. 3):

$$P = \frac{4.52}{C^{1.85} d^{4.87}} \times Q^{1.85}$$

where

p is the friction loss per foot of pipe in psi/ft, d is the internal diameter in inches, C is the Hazen-Williams coefficient, and Q is the flow in gpm

Thus, for any given segment of pipe, the friction loss is

$$\frac{4.52L}{C^{1.85} d^{4.87}} \times Q^{1.85}$$

where "L" is the length of the pipe, in feet.

$$\frac{4.52L}{C^{1.85} d^{4.87}}$$

can be calculated for the given segment of pipe and called "$k$" (with an identifying subscript) and the friction loss in that pipe segment is $k\,Q^{1.85}$.

Not knowing how the flow divides through a network, it is necessary to *ASSUME* a flow through each segment, $Q_A$. Calling the correct (unknown) flow $Q$, $Q = Q_A + \Delta$, where $\Delta$ is the correction that must be applied to $Q_A$. Students of math may vaguely recall something called the Binomial Theorem.

The Binomial Theorem can be expressed as follows:

$$(a+b)^n = a^n + na^{n-1}b + \frac{n(n-1)}{2!}a^{n-2}b^2 + \frac{n(n-1)(n-2)}{3!}a^{n-3}b^3 + \ldots nab^{n-1} + b^n$$

This looks rather formidable. And it is. But it is applicable to our problem. We have established that the friction loss in a pipe segment can be expressed as $k\,Q^{1.85}$

and that $Q = Q_A + \Delta$. Substituting for $Q$, the friction loss in the pipe segment becomes $k(Q_A + \Delta)^{1.85}$ and a relationship with the Binomial Theorem becomes apparent. Let $Q_A$, the assumed flow, be "$a$," $\Delta$, the difference between the assumed flow and the "correct" flow, be "$b$," and $n = 1.85$. For practical reasons we will discard all of the successive refinements to the right of the second term, "$na^{n-1}\,b$," in the Binomial Theorem and, substituting, we have

$$(Q_A + \Delta)^{1.85} = Q_A^{1.85} + 1.85Q_A^{1.85-1}\Delta$$

adding the "$k$," this becomes $k(Q_A + \Delta)^{1.85} = kQ_A^{1.85} + 1.85kQ_A^{1.85}\Delta$

It is time to take another look at our mini-grid.

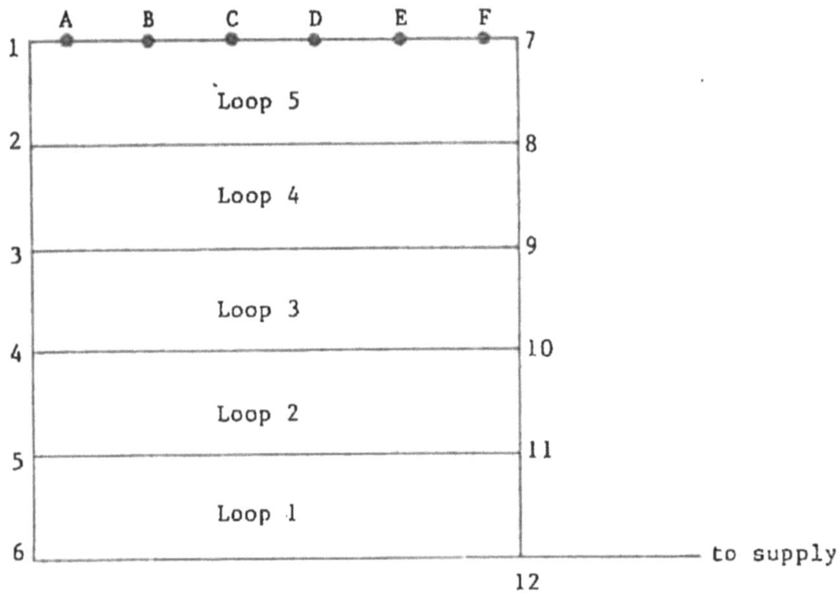


This consists of five loops. Look at loop 1. It consists of four legs, or pipe segments, 12–6, 6–5, 5–11, and 11–12. The friction loss in pipe segment 12–6 can be expressed as

$$k_{12-6}(Q_{A12-6} + \Delta)^{1.85}$$

and the friction loss in the other pipe segments can be expressed in a similar manner. We must now assume a direction of flow in each pipe segment and it seems reasonable to assume as follows:

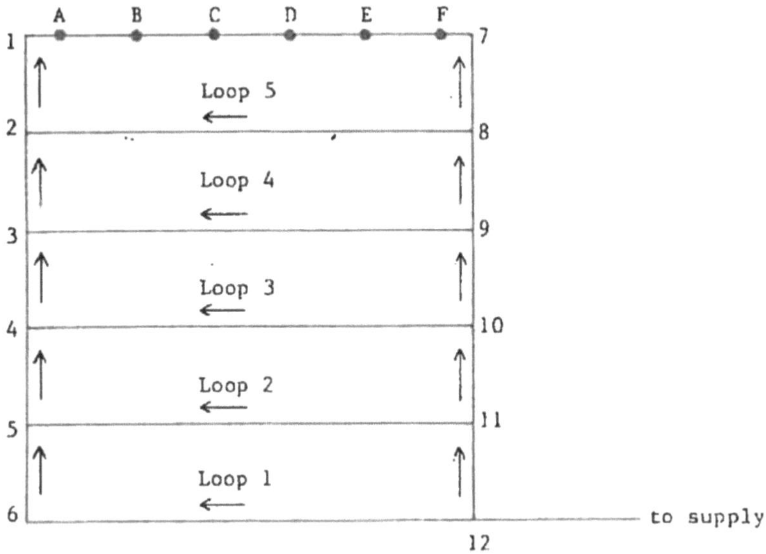


As a convention, we can consider the friction loss in all pipe segments flowing CLOCKWISE to be positive numbers and the friction loss in all pipe segments flowing COUNTERCLOCKWISE to be negative numbers. We will cloak this in more scholarly terms later but it should be self-evident that

$$k_{12-6}(Q_{A12-6}+\Delta)^{1.85}+k_{5-6}(Q_{A5-6}+\Delta)^{1.85}-k_{5-11}(Q_{A5-11}+\Delta)^{1.85}-k_{11-12}(Q_{A11-12}+\Delta)^{1.85}=0$$

In other words, the sum of the friction losses around a loop must equal zero. Look again at our abbreviated equation derived from the Binomial Theorem.

$$k(Q_A+\Delta)^{1.85}=kQ_A^{1.85}+1.85kQ_A^{1.85}\Delta$$

If we substitute the right side of the equation in our loop equation,

$$\Sigma\, kQ_A^{1.85}+1.85\Sigma\, kQ_A^{0.85}\Delta=0$$

Solving for $\Delta$,

$$1.85\Sigma\, kQ_A^{0.85}\Delta=-\Sigma\, kQ_A^{1.85}$$

$$\Delta=-\frac{\Sigma\, kQ_A^{1.85}}{1.85\Sigma\, kQ_A^{0.85}} \tag{18}$$

This is the Hardy Cross equation.

The mechanics of applying this equation can be best illustrated by an example. We will use our mini-grid and consider Sprinkler Heads C and D to be the flowing sprinklers. We must *assume* initial flows. For simplicity, we will assume all flow from Sprinkler Head C to be coming from the left and all flow from Sprinkler Head D to be coming from the right. We will assume the five non-discharging branch lines have an equal flow, 25 + 5 = 5 gpm. The example appears on the next page.

Note that all clockwise flows are positive numbers and all counterclockwise flows are negative numbers. While the $KQ^{1.85}$ value is positive or negative,

| Loop | Leg | $K$ | Assumed $Q_A$ | $KQ^{1.85}$ | $KQ^{0.85}$ | Δ Adjustment | New $Q_A$ |
|---|---|---|---|---|---|---|---|
| 1 | 12–6 | .0387516 | 5 | .761 | .152 | +.78 | +5.78 |
| | 6–5 | .0007596 | 5 | .015 | .003 | +.78 | +5.78 |
| | 5–11 | .0387516 | −5 | −.761 | .152 | +.78 − .26 | −4.48 |
| | 11–12 | .0004124 | −45 | −.472 | .010 | +.78 | −44.22 |
| | | | | Σ −.457 | Σ .317 | | |

$$\Delta = -\frac{\Sigma KQ^{1.85}}{1.85\Sigma KQ^{0.85}} = -\frac{-.457}{1.85 \times .317} = +.78$$

| | | | | | | | |
|---|---|---|---|---|---|---|---|
| 2 | 11–5 | .0387516 | 5 | .761 | .152 | +.26 − .78 | +4.48 |
| | 5–4 | .0007596 | 10 | .054 | .005 | +.26 | +10.26 |
| | 4–10 | .0387516 | −5 | −.761 | .152 | +.26 − .08 | −4.82 |
| | 10–11 | .0002250 | −40 | −.207 | .005 | +.26 | −39.74 |
| | | | | Σ −.153 | Σ .314 | | |

$$\Delta = -\frac{\Sigma KQ^{1.85}}{1.85\Sigma KQ^{0.85}} = -\frac{-.153}{1.85 \times .314} = +.26$$

| | | | | | | | |
|---|---|---|---|---|---|---|---|
| 3 | 4–10 | .0387516 | 5 | .761 | .152 | +.08 − .26 | +4.48 |
| | 4–3 | .0007596 | 15 | .114 | .008 | +.08 | +15.08 |
| | 3–9 | .0387516 | −5 | −.761 | .152 | +.08 + .12 | −4.80 |
| | 9–10 | .0002250 | −35 | −.162 | .005 | +.08 | −34.92 |
| | | | | Σ −.048 | Σ .317 | | |

$$\Delta = -\frac{\Sigma KQ^{1.85}}{1.85\Sigma KQ^{0.85}} = -\frac{-.048}{1.85 \times .317} = +.08$$

| | | | | | | | |
|---|---|---|---|---|---|---|---|
| 4 | 3–9 | .0387516 | 5 | .761 | .152 | −.12 − .08 | +4.80 |
| | 3–2 | .0007596 | 20 | .194 | .010 | −.12 | +19.88 |
| | 2–8 | .0387516 | −5 | −.761 | .152 | −.12 + .76 | −4.36 |
| | 8–9 | .0002250 | −30 | −.122 | .004 | −.12 | −30.12 |
| | | | | Σ +.072 | Σ .318 | | |

$$\Delta = -\frac{\Sigma KQ^{1.85}}{1.85\Sigma KQ^{0.85}} = -\frac{.072}{1.85 \times .318} = -.12$$

| | | | | | | | |
|---|---|---|---|---|---|---|---|
| 5 | 2–8 | .0387516 | 5 | .761 | .152 | −.76 + .12 | +4.36 |
| | 2–1 | .0007596 | 25 | .293 | .012 | −.76 | +24.24 |
| | 1–C | .0168264 | 25 | 6.489 | .260 | −.76 | +24.24 |
| | C–D | 0.005099 | 0 | — | — | −.76 | −.76 |
| | D–7 | .0168264 | −25 | −6.489 | .260 | −.76 | −25.76 |
| | 7–8 | .0002250 | −25 | −.087 | .003 | −.76 | −25.76 |
| | | | | +.967 | .687 | | |

$$\Delta = -\frac{\Sigma KQ^{1.85}}{1.85\Sigma KQ^{0.85}} = -\frac{967}{1.85 \times .687} = -.76$$

depending upon the sign assigned to $Q$, the $KQ^{0.85}$ value is always considered to be positive. Also note that, quite logically, the adjustment, $\Delta$, must be applied to the common legs in adjoining loops, and, when you do so, the sign of $\Delta$ is changed.

The single application of Hardy Cross adjustments in this example will never be sufficient in a grid. The same procedure must be repeated as many times as necessary to achieve a balance of flows within the desired tolerance. The "New Q" from this set of calculations becomes the "Assumed Q" for the next set of calculations. On the next page is a tabulation of the results of 22 additional Hardy Cross adjustments, continuing from the first adjustment.

You can see that even with this small grid, much smaller than the grids encountered in actual practice, the Hardy Cross calculations are very laborious if done manually on a hand calculator. So, having illustrated the method, we will forget about manual calculations.

There is one further point before leaving the subject, however. The Hardy Cross example is the traditional textbook method. There is no virtue in waiting until the next full pass to apply the common leg adjustment from the previous loop calculated. In fact, the "rippling" of the Hardy Cross adjustments is accelerated if the adjustment is made before calculating the adjustment for the next loop. The computer program is also simplified.

Till now we have ignored the question of which two sprinklers are hydraulically remote. Very arbitrarily, we will apply five Hardy Cross adjustments to our mini-grid, then calculate the friction loss from the sprinkler with the minimum pressure both ways around the perimeter of the grid to the point where the supply enters the grid. We will do this for each successive pair of sprinklers, starting with A-B until we see the "peaking," as NFPA 13 calls it. Here is the result:

| | A-B | B-C | C-D | D-E |
|---|---|---|---|---|
| FAR SIDE: | 6.19 | 7.45 | 7.86 | 7.46 |
| NEAR SIDE: | 7.02 | 7.84 | 7.94 | 7.29 |

It is very evident that C-D is the hydraulically remote pair of sprinklers. Perhaps we should look at a normal-size grid. A crude Hardy Cross program for the grid on Page 148 was run for three possible positions of the remote area, the area that the computer selected in the printout as the remote area plus the positions one sprinkler to the left and one sprinkler to the right. The first five Hardy Cross adjustments are shown.

## Successive Hardy Cross Adjustments to the Mini-Grid

| Leg | 0 | 1 | 2 | 3 | 4 | 5 | 6 | 7 | 8 | 9 | 10 | 11 | 12 | 13 | 14 | 15 | 16 | 17 | 18 | 19 | 20 | 21 | 22 | 23 |
|---|---|---|---|---|---|---|---|---|---|---|---|---|---|---|---|---|---|---|---|---|---|---|---|---|
| 12-6-5 | 5 | 5.78 | 5.89 | 6.07 | 6.11 | 6.15 | 6.17 | 6.19 | 6.2 | 6.21 | 6.22 | 6.23 | 6.23 | 6.23 | 6.24 | 6.24 | 6.24 | 6.24 | 6.24 | 6.24 | 6.24 | 6.24 | 6.24 | 6.24 |
| 5-11 | 5 | 4.46 | 4.81 | 4.71 | 4.79 | 4.78 | 4.81 | 4.82 | 4.83 | 4.83 | 4.84 | 4.85 | 4.85 | 4.85 | 4.86 | 4.86 | 4.86 | 4.86 | 4.86 | 4.86 | 4.86 | 4.86 | 4.86 | 4.86 |
| 11-12 | 45 | 44.22 | 44.11 | 43.93 | 43.89 | 43.85 | 43.63 | 43.8 | 43.79 | 43.76 | 43.77 | 43.77 | 43.77 | 43.76 | 43.76 | 43.76 | 43.76 | 43.76 | 43.76 | 43.76 | 43.76 | 43.76 | 43.76 | 43.76 |
| 5-4 | 10 | 10.25 | 10.69 | 10.78 | 10.9 | 10.94 | 10.99 | 11.01 | 11.03 | 11.05 | 11.06 | 11.07 | 11.08 | 11.09 | 11.1 | 11.1 | 11.1 | 11.1 | 11.1 | 11.1 | 11.11 | 11.11 | 11.11 | 11.11 |
| 4-10 | 5 | 4.82 | 4.46 | 4.42 | 4.36 | 4.37 | 4.35 | 4.36 | 4.35 | 4.36 | 4.36 | 4.36 | 4.36 | 4.36 | 4.36 | 4.36 | 4.36 | 4.36 | 4.36 | 4.36 | 4.36 | 4.36 | 4.36 | 4.36 |
| 10-11 | 40 | 39.74 | 39.3 | 39.22 | 39.1 | 39.06 | 39.01 | 38.99 | 38.97 | 38.95 | 38.94 | 38.93 | 38.92 | 38.91 | 38.9 | 38.9 | 38.9 | 38.9 | 38.9 | 38.9 | 38.9 | 38.9 | 38.9 | 38.9 |
| 4-3 | 15 | 15.08 | 15.15 | 15.2 | 15.25 | 15.3 | 15.34 | 15.37 | 15.39 | 15.41 | 15.42 | 15.44 | 15.44 | 15.45 | 15.45 | 15.46 | 15.46 | 15.47 | 15.47 | 15.47 | 15.47 | 15.47 | 15.47 | 15.47 |
| 3-9 | 5 | 4.8 | 4.39 | 4.35 | 4.3 | 4.27 | 4.26 | 4.24 | 4.24 | 4.23 | 4.23 | 4.22 | 4.22 | 4.22 | 4.22 | 4.22 | 4.22 | 4.22 | 4.22 | 4.22 | 4.22 | 4.22 | 4.22 | 4.22 |
| 9-10 | 35 | 34.92 | 34.85 | 34.8 | 34.75 | 34.7 | 34.67 | 34.63 | 34.61 | 34.59 | 34.58 | 34.56 | 34.56 | 34.55 | 34.55 | 24.54 | 34.54 | 34.53 | 34.53 | 34.53 | 34.53 | 34.53 | 34.53 | 34.53 |
| 3-2 | 20 | 19.88 | 19.54 | 19.56 | 19.55 | 19.57 | 19.6 | 19.61 | 19.63 | 19.64 | 19.65 | 19.66 | 19.67 | 19.67 | 19.67 | 19.68 | 19.68 | 19.68 | 19.68 | 19.68 | 19.69 | 19.69 | 19.69 | 19.69 |
| 2-6 | 5 | 4.36 | 4.66 | 4.57 | 4.59 | 4.56 | 4.54 | 4.53 | 4.51 | 4.51 | 4.5 | 4.49 | 4.49 | 4.49 | 4.48 | 4.48 | 4.48 | 4.48 | 4.48 | 4.48 | 4.47 | 4.47 | 4.47 | 4.47 |
| 8-9 | 30 | 30.12 | 30.46 | 30.44 | 30.45 | 30.43 | 30.4 | 30.39 | 30.37 | 30.36 | 30.35 | 30.34 | 30.33 | 30.33 | 30.32 | 30.32 | 30.32 | 30.32 | 30.32 | 30.32 | 30.31 | 30.31 | 30.31 | 30.31 |
| 2-1-C | 25 | 24.24 | 24.2 | 24.13 | 24.13 | 24.13 | 24.14 | 24.14 | 24.15 | 24.15 | 24.15 | 24.15 | 24.16 | 24.16 | 24.16 | 24.16 | 24.16 | 24.16 | 24.16 | 24.16 | 24.16 | 24.16 | 24.16 | 24.16 |
| C-D | 0 | 0.76 | 0.8 | 0.87 | 0.87 | 0.87 | 0.85 | 0.86 | 0.85 | 0.85 | 0.85 | 0.85 | 0.84 | 0.84 | 0.84 | 0.84 | 0.84 | 0.84 | 0.84 | 0.84 | 0.84 | 0.84 | 0.84 | 0.84 |
| D-7-8 | 25 | 25.76 | 25.8 | 25.87 | 25.87 | 25.87 | 25.86 | 25.86 | 25.85 | 25.85 | 25.85 | 25.85 | 25.84 | 25.84 | 25.84 | 25.84 | 25.84 | 25.84 | 25.84 | 25.84 | 25.84 | 25.84 | 25.84 | 25.84 |
| FAR P | | 7.78 | 7.8 | 7.83 | 7.85 | 7.86 | 7.87 | 7.88 | 7.89 | 7.9 | 7.9 | 7.9 | 7.91 | 7.91 | 7.91 | 7.91 | 7.91 | 7.91 | 7.91 | 7.91 | 7.91 | 7.91 | 7.92 | 7.92 |
| Near P | | 7.9 | 7.91 | 7.94 | 7.94 | 7.94 | 7.93 | 7.93 | 7.93 | 7.92 | 7.92 | 7.92 | 7.92 | 7.92 | 7.92 | 7.92 | 7.92 | 7.92 | 7.92 | 7.92 | 7.92 | 7.92 | 7.92 | 7.92 |

Note: "Far P" is the total friction loss from the sprinkler with the split flow, sprinkler C, around the far-side perimeter of the grid to the supply

"Near P" is the total friction loss from the sprinkler with the split flow, sprinkler C, around the near-side perimeter of the grid to the supply

Recognize that when the flows are properly balanced, the total friction loss from the point where the supply enters the grid to the sprinkler with the split flow will be the same by any path if all losses along the path in the direction of flow are considered positive numbers and all losses along the path against the direction of flow are considered negative numbers

| One sprinkler to the left: | 1 | 2 | 3 | 4 | 5 |
|---|---|---|---|---|---|
| Far side: | 35.71 | 36.35 | 36.87 | 37.51 | 38.07 |
| Near side: | 49.08 | 49.10 | 48.97 | 48.38 | 47.85 |
| Total: | 84.79 | 85.45 | 85.84 | 85.89 | 85.92 |

| Computer-selected remote area: | 1 | 2 | 3 | 4 | 5 |
|---|---|---|---|---|---|
| Far side: | 39.82 | 40.39 | 40.71 | 41.06 | 41.36 |
| Near Side: | 45.27 | 45.29 | 45.41 | 45.34 | 45.26 |
| Total: | 85.09 | 85.68 | 86.12 | 86.40 | 86.62 |

| One sprinkler to the right: | 1 | 2 | 3 | 4 | 5 |
|---|---|---|---|---|---|
| Far side: | 43.93 | 44.43 | 44.48 | 44.41 | 44.32 |
| Near side: | 41.15 | 41.19 | 41.51 | 41.88 | 42.20 |
| Total: | 85.08 | 85.62 | 85.99 | 86.29 | 86.52 |

The grid flows, in all cases, are still highly unbalanced after five adjustments but the hydraulically remote area can be inferred.

It is now time to dispense with mini-grids and look at the full-size grid to which we have alluded. On the following pages is the grid schematic, the computer printout of the calculation results and schematics showing the results. Unlike the majority of computer programs, this program numbers the legs, rather than the nodes.

SCHEMATIC OF A TYPICAL GRID

BRANCH LINE SPACING: 9.375'
16 BRANCH LINES, 1Ω " PIPE (SCHEDULE 40), INCLUDING RISER NIPPLES
15 SPRINKLER ON EACH BRANCH LINE. SPACED 10' APART LARGE ORIFICE SPRINKLER. k = 8.2
C = 120

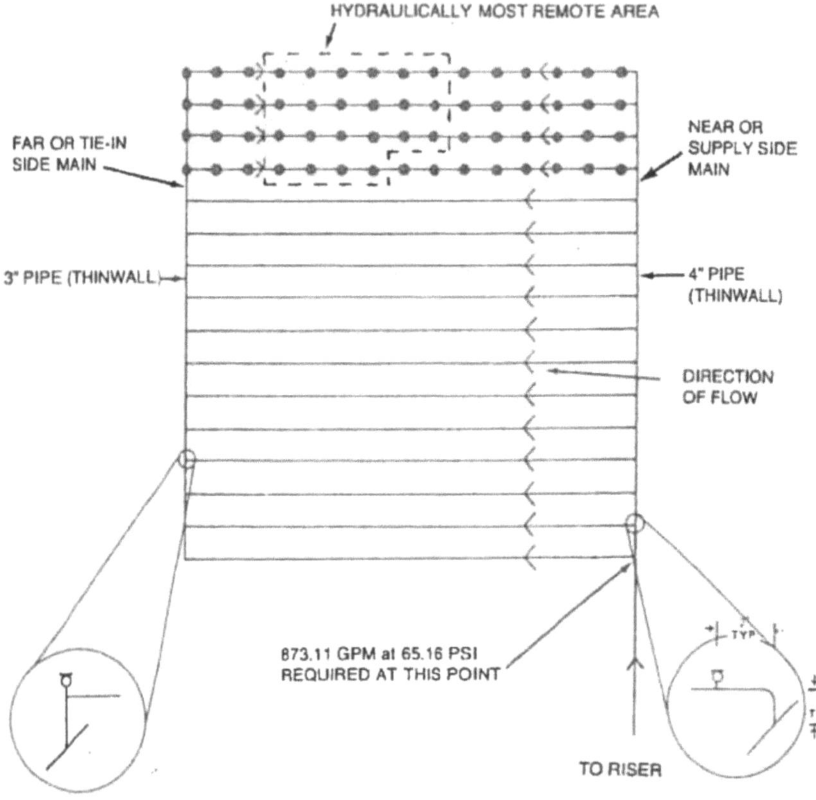


CALCULATE TO DELIVER A DENSITY OF .40 GPM PER SQUARE FOOT OVER 2000
SQUARE FEET AREA PER SPRINKLER = 10 x 9.375 = 93.75 SQUARE FEET
REQUIRED MINIMUM DISCHARGE PER SPRINKLER = .40 x 93.75 = 37.50 GPM

## COMPUTER PRINTOUT OF GRID CALCULATIONS

---- OUTER LOOP PRESSURE SUMMARY ----

OUTERLOOP (SUPPLY SIDE) SUMMARY

| LEG NO. | FLOW (GPM) | FRICTION LOSS | PRESSURE PSI |
|---|---|---|---|
| BEGIN | | | 20.91 |
| 5 | 36.59 | .49 | 21.41 |
| 6 | 74.53 | 1.84 | 23.25 |
| 7 | 114.07 | 30.36 | 53.61 |
| 39 | 114.07 | .04 | 53.65 |
| 40 | 228.15 | .14 | 53.79 |
| 41 | 342.22 | .30 | 54.08 |
| 42 | 429.06 | .45 | 54.53 |
| 43 | 466.03 | .53 | 55.06 |
| 44 | 500.93 | .60 | 55.66 |
| 45 | 534.58 | .69 | 56.34 |
| 46 | 567.59 | .76 | 57.09 |
| 47 | 600.56 | .84 | 57.93 |
| 48 | 634.08 | .93 | 58.86 |
| 49 | 668.68 | 1.02 | 59.89 |
| 50 | 704.87 | 1.13 | 61.02 |
| 51 | 743.07 | 1.25 | 62.26 |
| 52 | 783.65 | 1.37 | 63.64 |
| 53 | 826.92 | 1.52 | 65.16 |
| TOTAL (SUPPLY SIDE) | | 44.24 | 65.16 |

OUTERLOOP (TIE-IN SIDE) SUMMARY

| LEG NO. | FLOW (GPM) | FRICTION LOSS | PRESSURE PSI |
|---|---|---|---|
| BEGIN | | | 20.91 |
| 4 | .91 | .00 | 20.91 |
| 3 | 38.41 | .54 | 21.45 |
| 2 | 76.39 | 1.93 | 23.38 |
| 1 | 116.04 | 19.64 | 43.02 |
| 54 | 116.04 | .13 | 43.14 |
| 55 | 232.31 | .46 | 43.60 |
| 56 | 349.40 | .97 | 44.57 |
| 57 | 444.05 | 1.51 | 46.08 |
| 58 | 407.09 | 1.29 | 47.36 |
| 59 | 372.18 | 1.09 | 48.45 |
| 60 | 338.53 | .91 | 49.37 |
| 61 | 305.52 | .76 | 50.12 |
| 62 | 272.55 | .61 | 50.74 |
| 63 | 239.04 | .48 | 51.22 |
| 64 | 204.43 | .36 | 51.58 |
| 65 | 168.24 | .25 | 51.83 |
| 66 | 130.04 | .16 | 51.98 |
| 67 | 89.46 | .08 | 52.06 |
| 68 | 46.20 | .02 | 52.08 |
| 38 | 46.20 | 13.08 | 65.16 |
| TOTAL (TIE-IN SIDE) | | 44.24 | 65.16 |
| DIFFERENCES | | -.00 | -.00 |

93.75 SQFT/HEAD X .400 GPM/SQFT = 37.50 GPM/HEAD
37.50 GPM/HEAD WITH K= 8.20 20.91 PSI
LOSS THROUGH GRID: (SEE SHEET NO ) 44.24 PSI

----- GRID SYSTEM CALCULATION -----

NOTES:

(1) TOTAL FLOW THROUGH THE GRID = 873.11 GPM
(2) TOTAL LOSS THROUGH THE GRID = 44.24 PSI
(3) C-FACTOR USED = 120.0

| LEG NO | FLOW (GPM) | PIPE SIZE(I.D.) | TOTAL EQ LENGTH | FRICTION LOSS/FT | TOTAL LOSS THR. A LEG |
|---|---|---|---|---|---|
| 1 | 116.04 | 1.610 | 47.00 | .4178 | 19.64 |
| 2 | 76.39 | 1.610 | 10.00 | .1928 | 1.93 |
| 3 | 38.41 | 1.610 | 10.00 | .0540 | .54 |
| 4 | .91 | 1.610 | 10.00 | .0001 | .00 |
| 5 | 36.59 | 1.610 | 10.00 | .0494 | .49 |
| 6 | 74.53 | 1.610 | 10.00 | .1842 | 1.84 |
| 7 | 114.07 | 1.610 | 75.00 | .4048 | 30.36 |
| 8 | 116.27 | 1.610 | 47.00 | .4193 | 19.71 |
| 9 | 76.57 | 1.610 | 10.00 | .1936 | 1.94 |
| 10 | 38.55 | 1.610 | 10.00 | .0544 | .54 |
| 11 | 1.01 | 1.610 | 10.00 | .0001 | .00 |
| 12 | 36.53 | 1.610 | 10.00 | .0492 | .49 |
| 13 | 74.50 | 1.610 | 10.00 | .1840 | 1.84 |
| 14 | 114.07 | 1.610 | 75.00 | .4048 | 30.36 |
| 15 | 117.09 | 1.610 | 47.00 | .4248 | 19.97 |
| 16 | 77.23 | 1.610 | 10.00 | .1967 | 1.97 |
| 17 | 39.06 | 1.610 | 10.00 | .0557 | .56 |
| 18 | 1.39 | 1.610 | 10.00 | .0001 | .00 |
| 19 | 36.28 | 1.610 | 10.00 | .0486 | .49 |
| 20 | 74.38 | 1.610 | 10.00 | .1835 | 1.84 |
| 21 | 114.07 | 1.610 | 75.00 | .4048 | 30.36 |
| 22 | 94.64 | 1.610 | 47.00 | .2865 | 13.47 |
| 23 | 48.92 | 1.610 | 10.00 | .0845 | .85 |
| 24 | 3.81 | 1.610 | 10.00 | .0008 | .01 |
| 25 | 41.29 | 1.610 | 10.00 | .0618 | .62 |
| 26 | 86.84 | 1.610 | 95.00 | .2444 | 23.22 |
| 27 | 36.96 | 1.610 | 172.00 | .0503 | 8.66 |
| 28 | 34.90 | 1.610 | 172.00 | .0453 | 7.78 |
| 29 | 33.65 | 1.610 | 172.00 | .0423 | 7.28 |
| 30 | 33.01 | 1.610 | 172.00 | .0408 | 7.02 |
| 31 | 32.97 | 1.610 | 172.00 | .0407 | 7.01 |
| 32 | 33.52 | 1.610 | 172.00 | .0420 | 7.22 |
| 33 | 34.61 | 1.610 | 172.00 | .0446 | 7.66 |
| 34 | 36.19 | 1.610 | 172.00 | .0484 | 8.32 |
| 35 | 38.20 | 1.610 | 172.00 | .0535 | 9.20 |
| 36 | 40.58 | 1.610 | 172.00 | .0599 | 10.29 |
| 37 | 43.26 | 1.610 | 172.00 | .0673 | 11.58 |
| 38 | 46.20 | 1.610 | 172.00 | .0760 | 13.08 |
| 39 | 114.07 | 4.124 | 9.38 | .0041 | .04 |
| 40 | 228.15 | 4.124 | 9.38 | .0150 | .14 |
| 41 | 342.22 | 4.124 | 9.38 | .0317 | .30 |
| 42 | 429.06 | 4.124 | 9.38 | .0481 | .45 |
| 43 | 466.03 | 4.124 | 9.38 | .0561 | .53 |
| 44 | 500.93 | 4.124 | 9.38 | .0641 | .60 |
| 45 | 534.58 | 4.124 | 9.38 | .0723 | .68 |
| 46 | 567.59 | 4.124 | 9.38 | .0807 | .76 |
| 47 | 600.56 | 4.124 | 9.38 | .0896 | .84 |
| 48 | 634.08 | 4.124 | 9.38 | .0991 | .93 |
| 49 | 668.68 | 4.124 | 9.38 | .1093 | 1.02 |
| 50 | 704.87 | 4.124 | 9.38 | .1205 | 1.13 |
| 51 | 743.07 | 4.124 | 9.38 | .1329 | 1.25 |
| 52 | 783.65 | 4.124 | 9.38 | .1466 | 1.37 |
| 53 | 826.92 | 4.124 | 9.38 | .1619 | 1.52 |
| 54 | 116.04 | 3.260 | 9.38 | .0135 | .13 |
| 55 | 232.31 | 3.260 | 9.38 | .0486 | .46 |
| 56 | 349.40 | 3.260 | 9.38 | .1034 | .97 |
| 57 | 444.05 | 3.260 | 9.38 | .1611 | 1.51 |
| 58 | 407.09 | 3.260 | 9.38 | .1372 | 1.29 |
| 59 | 372.18 | 3.260 | 9.38 | .1162 | 1.09 |
| 60 | 339.53 | 3.260 | 9.38 | .0975 | .91 |
| 61 | 305.52 | 3.260 | 9.38 | .0807 | .76 |
| 62 | 272.55 | 3.260 | 9.38 | .0653 | .61 |
| 63 | 239.04 | 3.260 | 9.38 | .0512 | .48 |
| 64 | 204.43 | 3.260 | 9.38 | .0384 | .36 |
| 65 | 168.24 | 3.260 | 9.38 | .0267 | .25 |
| 66 | 130.04 | 3.260 | 9.38 | .0166 | .16 |
| 67 | 89.46 | 3.260 | 9.38 | .0083 | .08 |
| 68 | 46.20 | 3.260 | 9.38 | .0024 | .02 |

COMPUTER-ASSIGNED GRID LEG NUMBERS

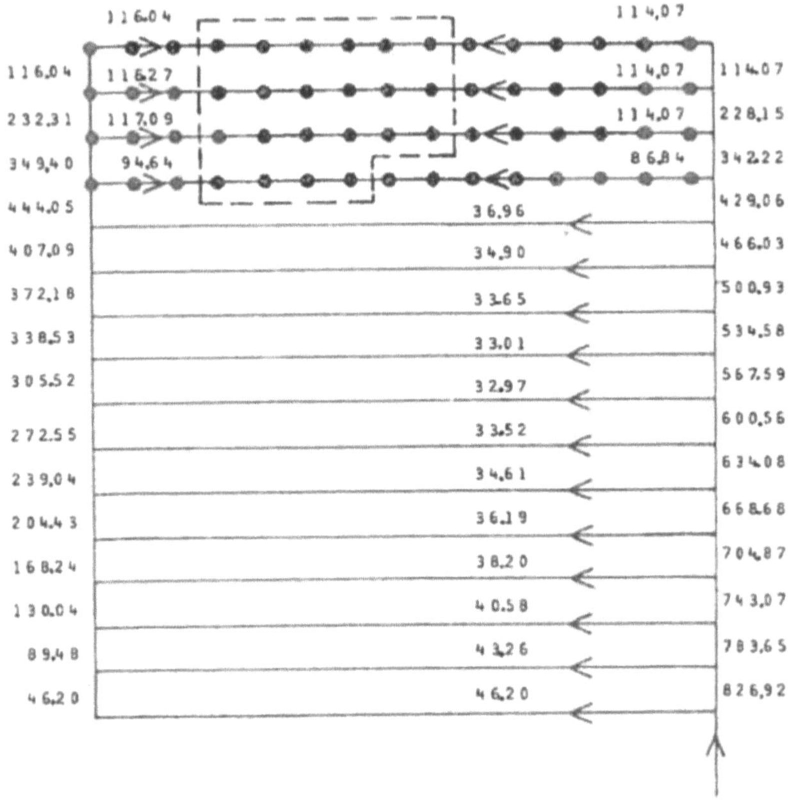


COMPUTER-CALCULATED FLOWS THROUGH EACH LEG OF GRID.

Seeming discrepancies in the second decimal place can be attributed to internal rounding-off. For example, 114.074 + 114.074 = 228.148. It this is rounded off to two decimal places, it becomes 114.07 + 114.07 = 228.15.

SCHEMATIC OF GRID OPERATING AREA

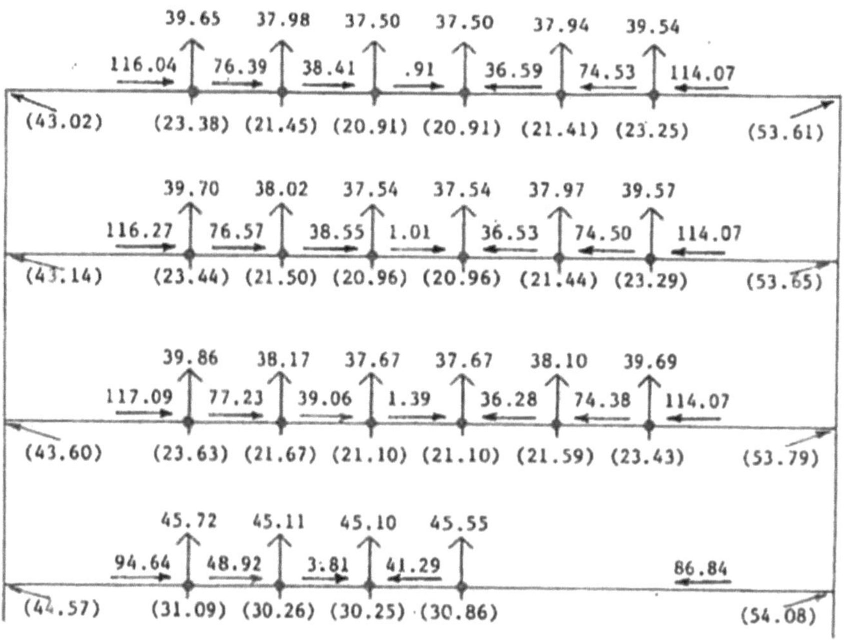


Schematic of Grid Operating Area

Numbers in parentheses represents pressure, in psi.
Numbers not in parentheses represent flow, in gpm.

Note that velocity pressure is not taken into account in these calculations. Computer programs for calculating grids normally ignore velocity pressure because of the extensive calculations involved. If velocity pressure were taken into account, the total flow would probably be reduced on the order of 13–15 gpm and there would be a small reduction in the required pressure. The pressures at the individual sprinkler would be higher and the governing head (the sprinkler with the minimum required flow of 37.50 gpm) would probably be the second sprinkler from the left rather than the fourth sprinkler from the left, and it would not necessarily be on the top branch line.

Reference is made elsewhere to "sprinklers where the flow splits." These are the fourth sprinkler from the left on the top three lines and the third sprinkler from the left on the bottom line.

Keeping in mind that you do not need a computer, or even a very powerful calculator, to check grid calculations, let us examine this printout.

The indicated flow in Leg 27, the first branch line below the design area, is 36.96 gpm. Referring to the supply side outer loop summary, the pressure at the right end of the branch line is shown as 54.53 psi. (The pressure is for the end of the leg toward the supply). Referring to the tie-in side outer loop summary, the pressure at the left end of the branch line is shown as 46.08 psi. Based upon the pressure differential, 54.53 -46.08 = 8.45 psi, what is the indicated flow? The friction loss constant for 1-½ in. Schedule 40 pipe is $6.330 \times 10^{-5}$ and the total equivalent length of the branch line is 172 feet.

$$8.45 = 172 \times 6.330 \times 10^{-5} Q^{1.85}$$
$$Q = 36.49$$

This differs significantly, 0.47 gpm, from the 36.96 shown in the printout. What are we to make of this? Is it acceptable? Let us look at the difference in friction loss.

$$172 \times 6.330 \times 10^{-5}(36.96)185 = 8.65\,\text{psi}$$

The discrepancy, in terms of pressure, is 8.65- 8.45 = 0.20 psi, well within the 0.50 psi balancing required by NFPA 13. As we have said, grid calculations are only approximations. All computer programs are designed to stop when specified tolerances have been attained. We have singled out the largest imbalance in this set of calculations. It is our opinion that these numbers are satisfactory. We have, incidentally, compared the results of this particular program with the results of some of the other commonly used programs and found that the numbers matched very closely. What is the "right" solution of this grid? We do not have the "right" answer but what follows is probably very close.

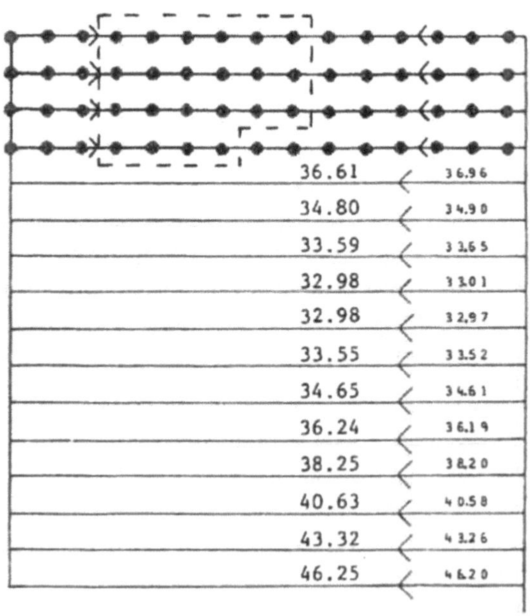


The large numbers are the "very close" flows, and the small numbers are the ones generated by the computer.

You will note that the "very close" flow in Leg 27, the first branch line below the operating area, is 36.61 gpm, compared to 36.96 gpm in the computer printout and 36.49 gpm based upon the pressure differential calculated by the computer. What really matters, however, is the bottom line. The friction loss through the grid on our "very close" solution is 44.26 psi, compared to 44.24 psi from the computer. The "very close" demand at the entrance to the loop is 873.12 gpm at 65.17 psi. The computer-generated 873.11 gpm at 65.16 psi is close indeed. Remember, of course, that while we should insist on a reasonably accurate computer output to avoid the unpredictable cumulative consequences of too many approximations, at best, everything to the right of the decimal point is meaningless. We are not even considering velocity pressure in these calculations.

When we demonstrated the use of the Hardy Cross method to calculate the mini-grid we were able to ignore the fact that we have pressure-dependent flows from the discharging sprinklers. Very conveniently, in an example on Page 146, the small flow (0.84 gpm) from discharging Sprinkler D to discharging Sprinkler C is on the order of 0.00046 psi, when converted to friction loss, not enough to increase the flow from Sprinkler C within two decimal places. With a normal grid having, perhaps, six or seven flowing sprinklers on a branch line, the flows from the sprinklers will change as the flow split is changed to achieve balancing. We do not intend to probe the occult art of programming grid calculations and will not attempt to determine the best way of handling this problem. In fact, there may not be a "best" way, and we need not be concerned with HOW the computer calculates the grid if the results will stand up to scrutiny.

One approach to simplifying iterative calculations would be to consider all flowing sprinklers on a line to be a point source or a single large sprinkler. This would give you a reasonable approximation that you could ultimately refine. Refer to the schematic of the grid operating area on Page 145. The total flow on the top branch line is 230.11 and the minimum, or split-sprinkler, pressure is 20.91. Applying the basic flow equation, this yields a "$k$" of 50.32. The flow from the left is 116.04 and the pressure differential between the split-flow sprinkler and the far-side main is 43.02 − 20.91 = 22.11 psi. Applying the equation

$$p = KLQ^{1.85}$$

where $p$ is the friction loss, $K$ is the friction loss constant from Table 4 of Appendix A, $L$ is the pipe length, and $Q$ is the flow,

$$L = \frac{P}{KQ^{1.85}} = \frac{22.11}{6.33 \times 10^{-5} \times 116.04^{1.85}} = 52.92\,\text{feet}$$

The actual equivalent pipe length from the first flowing sprinkler on the left side to the far main is 47.00 feet. Thus, the point source could be considered to be 5.92 feet to the right of that sprinkler. Similar calculations for the right side places the point source 5.79 feet to the left of the first flowing head on the right. Had it been assumed initially that the flow split evenly, the point source would have been placed about 5.84 feet in from both ends flowing sprinklers. Incidentally, if you think about it, you may realize that it is not merely good fortune that the flow splits very close to 50-50 in the hydraulic remote area. The area which splits closest to 50-50 can be expected to be the hydraulically most remote area.

While the remote flowing branch line is fairly simple, the other flowing branch lines present some problems. Looking again at our typical grid example and focusing on the second flowing branch line, it should be evident that the difference in the flows in this line, compared to the flows in the remote line, will be the consequence of the friction losses in the two side mains connecting them, Leg 54 and Leg 39. The incremental pressure is 0.13 on the left side and 0.04 on the right side, a total of 0.17 psi. This pushes our point source slightly to the right. The big problem, however, arises from the fact that the pressure at the sprinkler is a function of $Q^2$ and the friction loss in the line is a function of $Q^{1.85}$ It is possible, however, that a useful approximation could be obtained by ignoring the slight shift of the point source and by assuming a 50-50 flow split. Or, knowing that the flow split will be tilted toward the left, a slightly different ratio could be assumed.

The sum of the two incremental pressures, 0.17, is equal to the difference between the split-head pressures on the first and second lines plus the difference in the friction losses on the right and left sides in the first and second lines. If you formalize all of this into an equation, substitute the known numbers (with less rounding-off than the numbers above), and refine the equation, you will end up with

$$Q^2 + AQ^{1.85} = B$$

where $A$ and $B$ are known values. This is susceptible to an easy trial-and-error solution. You could then calculate the split-head pressure and allocate the left and right flows based upon the pressure differentials.

We will leave the subject of the mechanics of calculating a grid at this point. The people who have developed computer programs have their own proprietary secrets. Initial assumed flows can reflect an understanding of grid characteristics. There are techniques for accelerating convergence. More sophisticated mathematical techniques, developed for analyzing electrical circuits, are available, although we are told that they require considerable memory.

A loose analogy is sometimes made between hydraulics and electricity.

Flow (gpm) can be equated to current (amperes), friction loss (psi) to resistance (ohms), and pressure (psi) to voltage (volts). Kirchhoff's Laws, familiar to electrical engineers, can be applied analogously to a grid. Hydraulic calculations are more difficult, however, because in most electrical circuits the resistance is independent of the flow whereas friction loss varies exponentially with the flow.

In simple terms, Kirchhoff's Laws (Gustav Robert Kirchhoff, 1824–1887, German physicist) state that in a network of conductors, with sources of emf connected in one or more places, the distribution of currents must satisfy the following:

1. The algebraic sum of the currents toward any junction point is zero.
2. The algebraic sum of the voltages around any closed path in the network is zero.

Converting these two conditions into hydraulic terms and expressing them in simple English:

1. The flow into a junction point must equal the flow out of a junction point.
2. The algebraic sum of the friction losses in a loop is zero.

If somewhat self-evident, these are the basic elements of a grid program, although loops can be ignored. A grid can be viewed as having closed-end branch lines except that the precise location of the closed-ends is one of the unknowns. From this viewpoint, paths from the end of the closed-end branch lines to a common point in the supply can be analyzed.

We have already discussed more than anyone who works with hydraulically calculated sprinkler systems needs to know. It is time to return to the mundane matters that *really* matter.

# Personal Computer Programs for Hydraulic Calculations

Most of the commercially available computer programs for hydraulically calculating sprinkler systems share common characteristics and a general discussion of these characteristics may be useful.

All programs adhere to the NFPA 13-prescribed method for calculating friction loss. Therefore, all programs pretty much require the same input although the ways in which they ask for the information can differ significantly.

We have demonstrated the method for manual calculation of a tree system starting at the most remote sprinkler outlet and working your way through the system. The computer can be programmed to use a more sophisticated approach, solving the same equations, except solving all of the pieces simultaneously. Since all pieces of pipe are connected, there is the natural occurrence of equilibrium (conservation of energy and mass) within the piping system that can be used to solve for the friction loss in the network. So-called Newton–Raphson or complex matrix methods may be used. Some programs use loop-finding routines to determine the multiple paths through a piping network to the same endpoint. Such routines are transparent to the user.

We have already discussed a traditional method for solving looped piping configurations, the Hardy-Cross method. While it is a solid loop solution method, it requires user interaction for the defining of loop paths and it is very slow in its solution time when compared to other methods. If you are manually defining loop paths for looped and gridded systems, it indicates that your program uses Hardy-Cross solution methods.

Most hydraulic solution methods were developed as part of public university or federal research. Therefore, all of the pieces of information, except the user interface, necessary to develop a sprinkler hydraulics program are available through public and university libraries. For the normal user, however, this is not a practical alternative to purchasing a polished program.

Let us consider the input. Input data is the way in which you describe the piping network to the computer. Care must be taken with the input since, as they say, "Garbage IN, garbage OUT." Computers have no inherent common sense, although



H. S. Wass Jr. and R. P. Fleming P.E., *Sprinkler Hydraulics*,
https://doi.org/10.1007/978-3-030-02595-3_25

a well-designed program should not accept information that is not consistent with acceptable sprinkler system design. Granted, program developers properly view the program as a tool, not a substitute for engineering judgment. This view does not, however, preclude having the program questioning a significant deviation from the norm to verify that the entry is intentional, not accidental. Input data can be broken down into four simple groups:

1. Nodes.
2. Pipes.
3. Sources.
4. Pumps.

Nodes identify points in the piping network which require an adjustment to the Hazen–Williams friction loss formula; that is, a change in the “d”, the “Q”, or the “C”, and, in the case of a booster pump, a change in the “p.” Thus nodes are needed at the following locations:

1. Change in internal pipe diameter.
2. Change in the Hazen–Williams C-value (Normally this change would not occur within the sprinkler system but will occur when you get to the underground supply.)
3. Flow splits within the piping network.
4. Sprinklers assumed to be discharging.
5. Hose connections assumed to be discharging.
6. Hydrants assumed to be discharging.
7. Booster pump inlet and discharge.
8. Check valves, pressure regulating valves, and fixed or flow-dependent pressure loss devices.

Too many nodes are better than too few, but all nodes must be located properly.

Piping information must be provided for every piece of pipe in the system as follows:

1. Nodes connected by the pipe; that is, identify the “begin” node and “end” node.
2. Centerline pipe length from the “begin” node to the “end” node.
3. Internal pipe diameter. While the computer must use the actual internal pipe diameter, it may be sufficient to input the nominal pipe diameter if the type of pipe has been identified, enabling the computer to obtain the actual internal diameter from a table.
4. Total equivalent fitting length (from “begin” node to “end” node). Again, it may be sufficient to simply input the fitting, such as an elbow, and let the computer look up the appropriate equivalent pipe length.
5. Hazen– Williams C-value for the pipe segment. This value should be established at the outset and assumed for each segment of pipe until told otherwise.

In addition to the actual piping information, various types of system hardware should be input, when present, such as:

1. Flow-dependent pressure loss devices (backflow preventers).
2. Fixed pressure loss devices.
3. Pressure regulating devices.
4. Check valves.

Some programs may address all of these devices, while others attempt only a few. As we have discussed elsewhere, backflow preventers present a particular challenge because the associated pressure loss varies with the flow in a manner unique to each make and model and is not reducible to a simple equation. Since the pressure loss is likely to be very significant, they should not be ignored but not all programs address them.

All programs require the definition of a water supply. Two types of water supplies can normally be defined within a sprinkler program—a municipal supply or a static reservoir with a fire pump. The municipal supply can be defined by a hydrant flow test, while the static supply may be defined by the characteristics of the pump or pumps.

There are two types of sprinkler system calculations that can be performed. The most common type, sometimes called a "Minimum Pressure" or "Demand" calculation, starts with the minimum flow or pressure required to meet the desired design criteria and calculates the friction loss from the hydraulically remote area to the water source. The calculated flow and pressure is then compared to the available flow and pressure at the water source. Unless, perchance, the calculated required flow and pressure matches the water supply, changes can be made in the pipe sizing. The final calculated flow and pressure must be available from the water supply but economic considerations usually eliminate any significant cushion in the water supply.

A second type of calculation is sometimes referred to as a "Maximum Pressure" or "Supply" calculation. This calculation uses the water supply to determine the actual flow and pressure at any point in the system with the specified sprinklers flowing and a specified hose demand.

The "Minimum Pressure" calculation is the common method used when submitting sprinkler system calculations in accordance with NFPA 13 requirements. The "Maximum Pressure" calculations are useful to engineering and insurance people to evaluate the capability of the sprinkler system to meet the sprinkler demand imposed by the current occupancy.

While the required data that must be entered is fairly consistent between programs, the method, or interface, used by the programmer to get the information is not. The method of entering the data depends upon the programmer's expertise, the programming language, and the developer's perception of what is easy for the user. There are programs that are broken into three parts—an entry program, a calculation program, and a review program. This can be very awkward.

The buzzword of the late 1980s was "spreadsheet editing." This was a phrase coined by B. W. Melly in his review of then-current hydraulic programs in 1988. Spreadsheet editing allows the user to move freely about an editing screen, broken into predefined rows and columns, making data entry, review, and editing more organized. This format, not frequent then, has become an industry standard since the review was published.

The real challenge for the program developer is to provide a user entry/editing routine that is easy and self-explanatory. It should not require frequent reference to a user manual. Of course, the program must be written for a wide range of users, both in terms of computer expertise and sprinkler hydraulics knowledge. The program should require only the most basic knowledge of computers. A knowledge of the physical elements of a sprinkler system must be assumed but there is technically no need for the user to have ever heard of the Hazen–Williams equation. In that vein, the needs of the novice should be met with adequate guidance on the input of the water supply since the simple entry of a hydrant flow test, however valid the test, could lead to very erroneous results.

A program can take things only so far and thorough supporting documentation and example problems are needed to assist the user but the program developer should have a conscious goal of minimizing the occasions when the user will have to refer to the documentation. The developer must also be aware that the program's use becomes second nature as it is being developed and what seems like the quick, easy, and obvious way to work may not strike anyone else that way.

The NFPA 13 annex also includes an example of the calculation of a tree system which dates back to when hydraulic calculations were first introduced into the standard. This was before the advent of personal computers and the output format was designed for hand calculations (with the aid of a calculator), which were common to that era. Since hand calculations are no longer the norm, a different format would be appropriate to the computer-generated data.

The following two pages show an example of a "stacked" computer-generated printout with three lines per pipe to fit on a standard sheet of paper. It is based on the tree system from the NFPA 13 annex, but with a small difference in numbers attributable to rounding errors.

Beginning with the 2010 Edition of NFPA 13, rules were inserted to standardize the formats of computer-generated hydraulic reports. The rules require a summary sheet, a graph sheet, a water supply analysis, a node analysis, and detailed worksheets, presented in that specific order. The intent was to make it easier for municipal officials and other authorities having jurisdiction to properly review the hydraulic calculation submittals.

Computer-Aided Drafting and Design (CADD) programs are now being used by most sprinkler contractors to prepare their working drawings. The transition from manual drafting to CADD has been more of a leap than a simple step. The guidelines of the CADD design software must be rigorously followed. There is still the need to specify exact pipe lengths, pipe diameters, and fitting dimensions. Screen color must be standardized since it is directly tied to the plotter pen type or line thickness generated for the output device. Since the drawings are done in the plan view, the designer must account for the vertical sprigs and drops, which do not have a defined line on the drawing, with descriptions on the drawings. A tee's dimensions must be placed at the intersection of the perpendicular lines on the drawing. While well-designed software can ease the process of adapting to computer design, the day is still some time off before the computer will not be dependent upon a knowledgeable sprinkler designer.

Ultimately, one might suppose that the sprinkler system design technician will be eliminated. The computer will be "taught" all there is to know about NFPA 13 and will be supplied with current pricing information. The computer-generated building plans will then be used to design the sprinkler system. Even then there will be a need for individuals with proper training, experience, and common sense to evaluate the proposed system design to ensure that it will provide appropriate protection for the intended application.

# Checklist for Reviewing Sprinkler Calculations

Most hydraulic calculations being made today utilize one of a number of commercially available programs. The output from these programs differs in appearance but will normally be adequate, especially since they are required to follow the NFPA 13 hydraulic report format introduced in the 2010 Edition. You can probably assume that the program performed the calculations correctly. Errors normally arise from incorrect input to the program.

1. Determine the appropriate design criteria for the occupancy in terms of the density and area along with the appropriate hose stream allowance. Some, but by no means all, of the important items that may be critical in determining the appropriate sprinkler design criteria are occupancy, NFPA occupancy classification group, NFPA commodity classification, use of encapsulation, storage arrangement (piles, single, double or multi-row racks, etc.), height, ceiling/roof height, clearance from top of storage to sprinkler deflectors, temperature rating of sprinklers, and type of sprinkler system (wet, dry).
2. Determine the manufacturer, model number, orifice size, and temperature rating of the sprinklers being used. Verify the "$k$" for the sprinklers and ascertain that the sprinkler is appropriate for this application. If it is a specially listed sprinkler, be aware of the listing requirements.
3. Determine the kind or kinds of pipe. This information is sometimes omitted from the plans but is essential for confirming that the proper internal diameters are being used in the calculations.
4. You may wish to highlight the hydraulic reference points on the plans. Reference points normally are required as follows:

   - At each sprinkler assumed to be flowing in the design area.
   - At all fittings where flow splits occur, such as the junction of a flowing branch line and across main.
   - At all points where the internal pipe diameter changes (Exception: dead-end lines outside of the design area.).



H. S. Wass Jr. and R. P. Fleming P.E., *Sprinkler Hydraulics*,
https://doi.org/10.1007/978-3-030-02595-3_26

This could also include a change of C-value if it is the connection to the underground.
- At all points where a change of elevation occurs.
- On the suction and discharge sides of fire pumps and points of connection where there are multiple water sources.

5. Having noted the sprinklers that are assumed to be flowing, ascertain that the appropriate sprinklers have been chosen. Keep in mind the considerations in determining the hydraulically remote area and the elongation of this area along the branch lines. Be particularly cautious where the hydraulically remote area may be problematic, such as a highly unsymmetrical system that might be encountered in a hospital. Carefully evaluate a design for an area that is not the hydraulically remote area. An example of this might be the protection of a rack storage area that is not in the remote area of the system.
6. Carefully check the distance between sprinklers, the distance between branch lines and the resultant area "covered" per sprinkler. Does the area conform to NFPA 13 or other applicable NFPA standards? If it is a specially listed sprinkler, does it conform to the listing? Beware of the unsymmetrical system. Sometimes the area per sprinkler in the remote area is less than the area per sprinkler in other areas and it is possible that, while the density will be met in the remote area, it may not be met in another area. Verify that every sprinkler in the operating area is discharging the required minimum flow based upon the density and the area "covered."
7. Carefully check the basic input as it is reflected in the computer printout.

   - Do all pipe sizes and pipe lengths agree with the plan?
   - Have all fittings been included?
   - Have they used the correct sprinkler "*k*"?
   - Is the correct Hazen-Williams "*C*" being used?
   - Where calculated "*k*"'s are used for sprinklers on riser nipples, armovers, or for branch lines, verify the "*k*". Also, be sure that they haven't taken a "one *k* fits all" approach when there are variations in the configuration for which the "*k*" has been calculated.
   - Are all elevation changes correctly included?

8. Verify that the indicated discharge from the sprinklers corresponds with the indicated pressure at each sprinkler. Determine that there is consistency in the use of, or omission of, velocity pressure.
9. Check that the total flow is equal to the sum of the individual flows from the sprinklers in the design area.
10. Verity that the sum of the flows entering a junction point is equal to the sum of the flows leaving the junction point. If it is a grid, slight discrepancies are acceptable. How much is slight is up to the judgment of the reviewer.
11. Ceiling sprinkler demand, in-rack sprinkler demand, if any, and the hose stream allowance should be combined for comparison to the water supply. When there are in-rack sprinklers, the in-rack demand must be properly balanced with the

ceiling demand at the actual point of connection. If there is an inside hose demand, ascertain that it has been added in the correct amount at the right point.

12. Sprinkler demand and water supply calculations must be adjusted to a common point for comparison.
13. FOR GRIDS ONLY: Check the pressure tolerances. The computer printout will normally show the pressure at each point in the two side mains where branch lines connect, with this pressure based upon the friction loss calculated through each segment of the side mains. In addition, most printouts will list a separate set of friction loss calculations based upon the flow through each branch line. Make a spot check of how well these pressures match up. NFPA permits a tolerance of 0.50 psi. With a good computer program, the differences should be somewhat less.

# In-rack Sprinkler Design

In-rack sprinklers are occasionally used to protect rack storage, and are an essential part of some protection for options within NFPA 13. When they are optional they are still highly desirable since they are likely to lead to faster control of fire and less damage to the stored commodity. Aside from the added expense and less flexible storage arrangement, the big concern with in-rack sprinklers is mechanical damage, leading to water damage, from careless forklift operators. This is one of the reasons in-rack sprinklers are required to have a separate indicating control valve.

Following are the necessary criteria for an in-rack sprinkler installation:

1. Sprinkler spacing and location, both vertically and horizontally, within the racks.
2. Minimum sprinkler discharge pressure or flow. NFPA 13 traditionally specified minimum 15 psi for in-rack sprinklers for storage to 25 feet and in some protection options specifies minimum flows per sprinkler, up to 120 gpm per sprinkler when ESFR sprinklers are used for in-rack protection schemes that are independent of ceiling sprinkler design.
3. The number of hydraulically most remote sprinklers assumed to be operating and on what levels.

NFPA 13 has traditionally stated:

> Water demand of sprinklers installed in racks shall be added to ceiling sprinkler water demand at the point of connection. Demands shall be balanced to the higher pressure.

While it should be noted that there may be some new special protection schemes that do not require simultaneous calculation of ceiling and in-rack sprinkler demands, most do.

"Balancing to the higher pressure" is necessary to determine the "real-life" flow that would occur if all of the sprinklers in the assumed ceiling operating area and the sprinklers assumed to be operating on the in-rack sprinkler lines were open and the minimum criteria for both ceiling and in-rack sprinklers are being met. This means that when the in-rack sprinklers are calculated (after calculating the ceiling sprinklers), you start with the minimum required in-rack sprinkler discharge



H. S. Wass Jr. and R. P. Fleming P.E., *Sprinkler Hydraulics*,
https://doi.org/10.1007/978-3-030-02595-3_27

pressure and work back to the junction point with the main supplying the ceiling sprinklers.

Unless through great luck the in-rack sprinkler pressure at the junction point equals the required ceiling pressure within an acceptable tolerance, perhaps 0.5 psi, adjust the flow associated with the lower pressure to the higher pressure, combine the adjusted flow with the other flow and carry that flow at the higher pressure back to the base of the riser.

In the more likely event that the pressures do not match, there are several choices. The calculated flow to the area producing the lower pressure can be increased until the two pressures match, the pipe supplying the area producing the lower pressure can be reduced in size until the pressures match, or a combination of both can be done. Conversely, the pipe sizing to the area producing the higher pressure could be increased to eliminate the difference. In any case, the total design must meet the requirements imposed by the available water supply and, in practice, this may foreclose some of the options listed above.

A special case arises when in-rack sprinklers are added when there is an existing sprinkler system. The ceiling system (which may, in fact, be a pipe schedule system that has never been calculated for anything) must be calculated for an appropriate density and area in conjunction with the in-rack sprinklers. The resultant pressure required for the ceiling sprinklers at "the point of connection" of the in-rack sprinklers could be specified as a fixed end-point for the in-rack sprinkler hydraulic calculations. We say "could be specified." What should be specified may require some judgment. We will leave the judgment to you, but a look at the hydraulics is in order.

Assume rack storage requiring a density of 0.18 gpm per sq. ft. over 2000 sq. ft. with one level of in-rack sprinklers where NFPA 13 specifies that the in-rack water demand be based on simultaneous operation of the 6 most hydraulically remote sprinklers with a minimum discharge pressure of 15 psi. The in-rack system supply is taken off at the base of the riser of the ceiling system.

Assume that the existing ceiling sprinkler system is calculated and requires 450 gpm at 65 psi at the base of the riser, including 9 psi for elevation. It is now necessary to estimate the flow from the 6 in-rack sprinklers. 160 gpm might be a reasonable guess. NFPA 13 also specifies a 500 gpm hose-stream allowance. Let us plot all of this, as in the graph at the top of page 166.

For future use, we have also plotted the estimated hydraulic characteristics of the in-rack system with an estimated elevation. For clarity, the flow has been multiplied by 5.

In the unlikely event that the water supply matches the demand, as in the graph at the bottom of page 166, it is all very simple and we specify a pressure of 65 psi for the in-rack system at the point of connection at the base of the riser.

What do we specify in the happy event that the water supply exceeds the demand, as in the graph at the top of page 166?

We could specify the same 65 psi. This would permit substantial deterioration of the water supply without compromising our minimum ceiling and in-rack criteria. What, then, would be the "real-life" situation now?

By trial and error, it can be determined that, with the remote 2000 sq. ft., ceiling area flowing, with the remote 6 in-rack heads flowing, and 500 gpm being used by the fire department for hose streams, the pressure at the junction point of the in-rack supply and the ceiling supply would be about 82 psi. The in-rack flow would be about 184 gpm (about 920 on the "5X" curve + 5) and the ceiling density would be about (530 + 450) × 0.18 = 0.21 gpm per sq. ft. This is illustrated in the graph at the bottom of page 167.

Note one thing in particular. If we specified balancing the in-rack and ceiling systems at 82 psi; that is, balancing the estimated 160 gpm in-rack demand, we would be reducing the in-rack demand by just a bit over 20 gpm, which will increase the supply to the ceiling sprinklers by a lesser amount, as you will see if you plot it out. Thus, the balance point is usually not critical. Nevertheless, we do not think it should be ignored.

To illustrate the "balancing" considerations, there is an example on pages 168 and 169. Assume that in-rack sprinklers are installed in the building with the pitched roof for which the ceiling sprinkler system was calculated on pages 97 through 100. Assume rack storage of a Class III commodity, with NFPA 13 specifying that the water demand for in-rack sprinklers should be based upon the operation of the most hydraulically remote six sprinklers when one level of in-racks is used, with a minimum discharge pressure of 15 psi. Understand that this should be in conjunction with the ceiling sprinkler design area. We will assume 8-ft. aisles between the racks, which means that the maximum permissible spacing between the in-rack heads is 12 feet and we will utilize this maximum spacing.

Referring to the example on page 169, the six in-rack sprinklers must flow 163.2 gpm and the pressure at the first head on the line is 53.49 psi. From this first sprinkler to the base of the riser there are 8 + 5 + 130 = 143 feet of pipe plus the *equivalent pipe length* of 7 elbows, 1 tee, and a gate valve. We will now review the consequences of three options, 2-in., 2-in., and 3-in. pipe from the first head to the point of connection with the main riser at the base of the riser.

2-in. pipe:

$$\text{Total equivalent pipe length} = 143 + 7 \times 5 + 10 + 1 = 189$$

$$\text{Friction loss} = 0.2325(\text{per foot}) \times 189 = 43.94$$

$$\text{Elevation} = 10 \times 0.433 = 4.33$$

$$\text{Total pressure} = 4.33 + 43.94 + 53.49 = 101.76\,\text{psi}$$

2½-in. pipe:

$$\text{Total equivalent pipe length} = 143 + 7 \times 6 + 12 + 1 = 198$$

$$\text{Friction loss} = 0.0979(\text{per foot}) \times 198 = 19.3$$

$$\text{Elevation} = 10 \times 0.433 = 4.33$$

$$\text{Total pressure} = 4.33 + 19.38 + 53.49 = 77.20\,\text{psi}$$

3-in. pipe:

$$\text{Total equivalent pipe length} = 143 + 7 \times 7 + 15 + 1 = 208$$
$$\text{Friction loss} = 0.0340(\text{per foot}) \times 208 = 7.07$$
$$\text{Elevation} = 10 \times 0.433 = 4.33$$
$$\text{Total pressure} = 4.33 + 7.07 + 53.49 = 64.89\,\text{psi}$$

The base of riser pressure for the ceiling system was calculated to be 75.54 psi and the associated flow for the ceiling system was determined to be 579.0 gpm. It is now necessary to balance the ceiling and the in-rack flows to the higher pressure.

2-in. pipe: The in-rack pressure of 101.76 governs and the ceiling flow must be adjusted

$$579.0\left(\frac{101.76 - 10.10}{75.54 - 10.10}\right)^{0.54} = 694.5\,\text{gpm}$$

Total demand = 694.5 + 163.2 = 857.7 gpm @ 101.74 psi

2½-in. pipe: The in-rack pressure of 77.20 psi governs and the ceiling flow must be adjusted

$$579.0\left(\frac{77.20 - 10.10}{75.54 - 10.10}\right)^{0.54} = 586.9\,\text{gpm}$$

Total demand = 586.9 + 163.2 = 750.1 gpm @ 77.18 psi

3-in. pipe: The ceiling pressure of 75.54 governs and the in-rack flow must be adjusted

$$163.2\left(\frac{75.54 - 4.33}{64.89 - 4.33}\right)^{0.54} = 178.2\,\text{gpm}$$

Total demand = 579.0 + 178.2 = 757.2 gpm @ 75.54 psi

In this example, the 2-in. pipe is clearly too small. There is very little to choose between the 2-in. and 3-in. pipe, and the 2-in. pipe would probably be the choice for economic reasons. It should be evident that 3-in. pipe would be undesirable because the in-rack demand would be increased as a result of the reduced friction loss.

You will note that the elevation pressure is being deducted in the balancing calculations. The 4.33 psi is for the in-racks and the 10.1 psi represents the sum of the elevation corrections for the ceiling system. In the "3-in. pipe," the adjustment is being made along the in-rack curve whereas the others are along the ceiling-system curve. The use of the 0.54 power is only a reasonable approximation, as discussed elsewhere.

All things considered, despite the length of the treatment, the "balancing act" is one of the less important parts of this book. But the credibility of calculations requires attention to detail since what is important and what is not important are usually not obvious.

In some cases where a 30 psi end-sprinkler pressure is specified for the in-rack sprinklers, the resultant pressure for the in-rack system can pose a design problem and the use of larger orifice sprinklers with a minimum pressure of 15 psi is acceptable. The flow from a K-8.0 large orifice sprinkler at 15 psi is virtually identical to the flow from a standard K-5.6 orifice sprinkler at 30 psi.

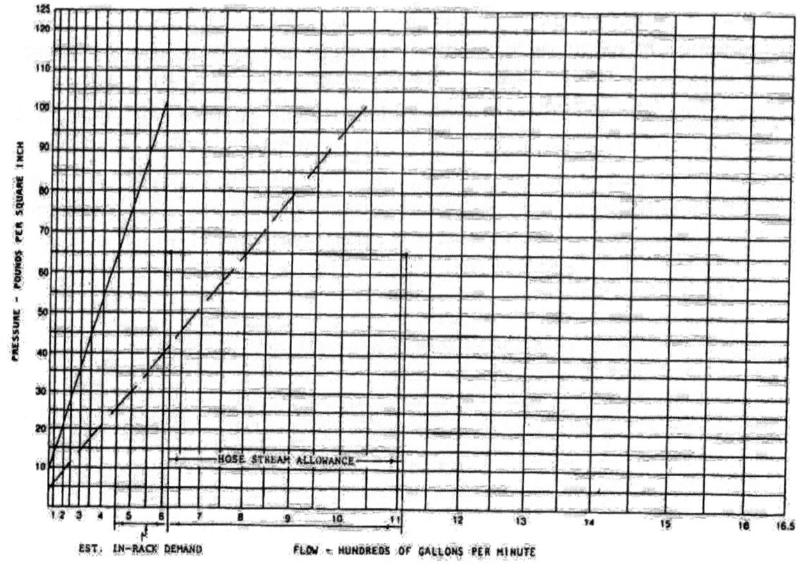
PRESSURE - POUNDS PER SQUARE INCH
125
120
110
100
90
80
70
60
50
40
30
20
10
HOSE STREAM ALLOWANCE
1 2 3 4 5 6 7 8 9 10 11 12 13 14 15 16 16.5
EST. IN-RACK DEMAND
FLOW - HUNDREDS OF GALLONS PER MINUTE

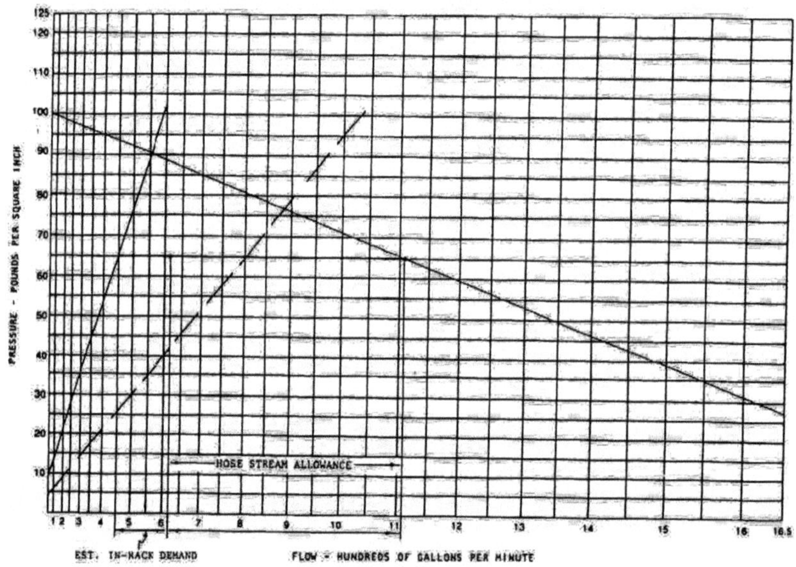
PRESSURE - POUNDS PER SQUARE INCH
125
120
110
100
90
80
70
60
50
40
30
20
10
HOSE STREAM ALLOWANCE
1 2 3 4 5 6 7 8 9 10 11 12 13 14 15 16 16.5
EST. IN-RACK DEMAND
FLOW - HUNDREDS OF GALLONS PER MINUTE

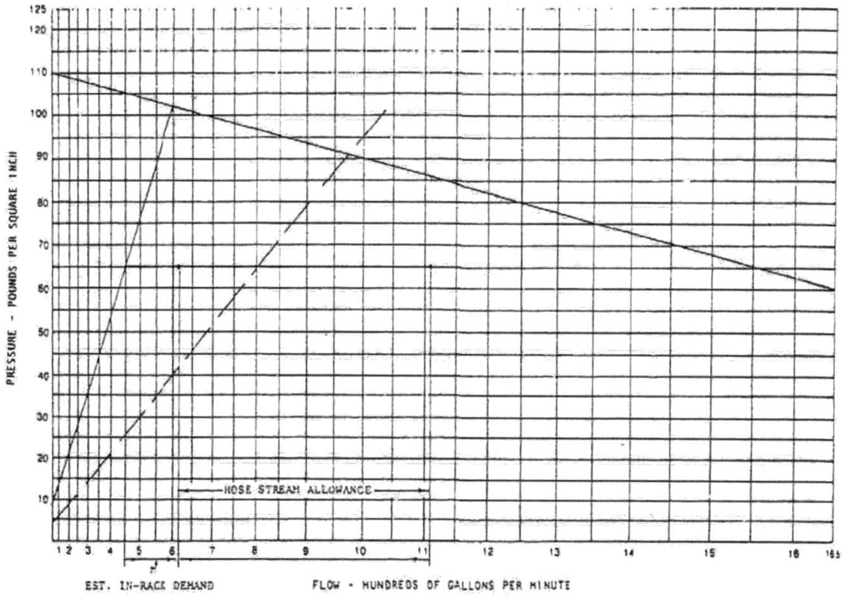
PRESSURE - POUNDS PER SQUARE INCH
HOSE STREAM ALLOWANCE
EST. IN-RACK DEMAND
FLOW - HUNDREDS OF GALLONS PER MINUTE

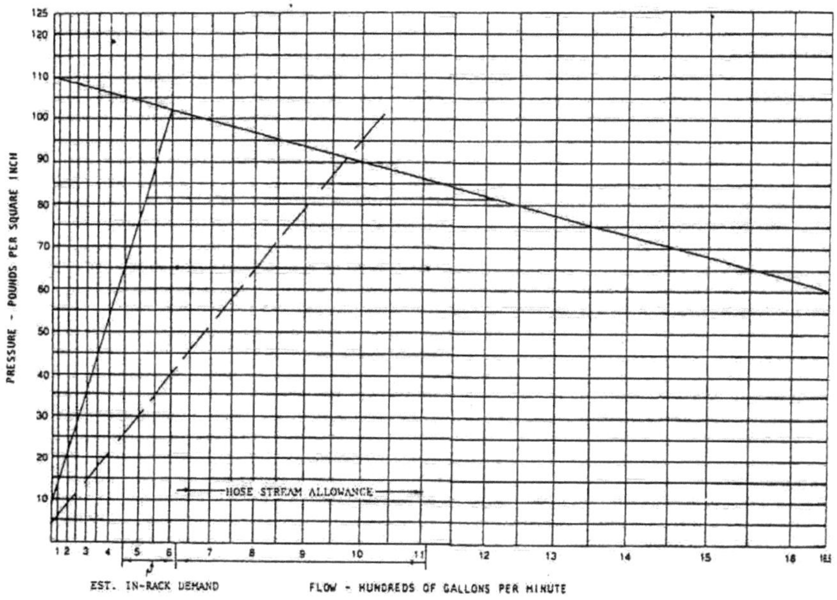
PRESSURE - POUNDS PER SQUARE INCH
HOSE STREAM ALLOWANCE
EST. IN-RACK DEMAND
FLOW - HUNDREDS OF GALLONS PER MINUTE

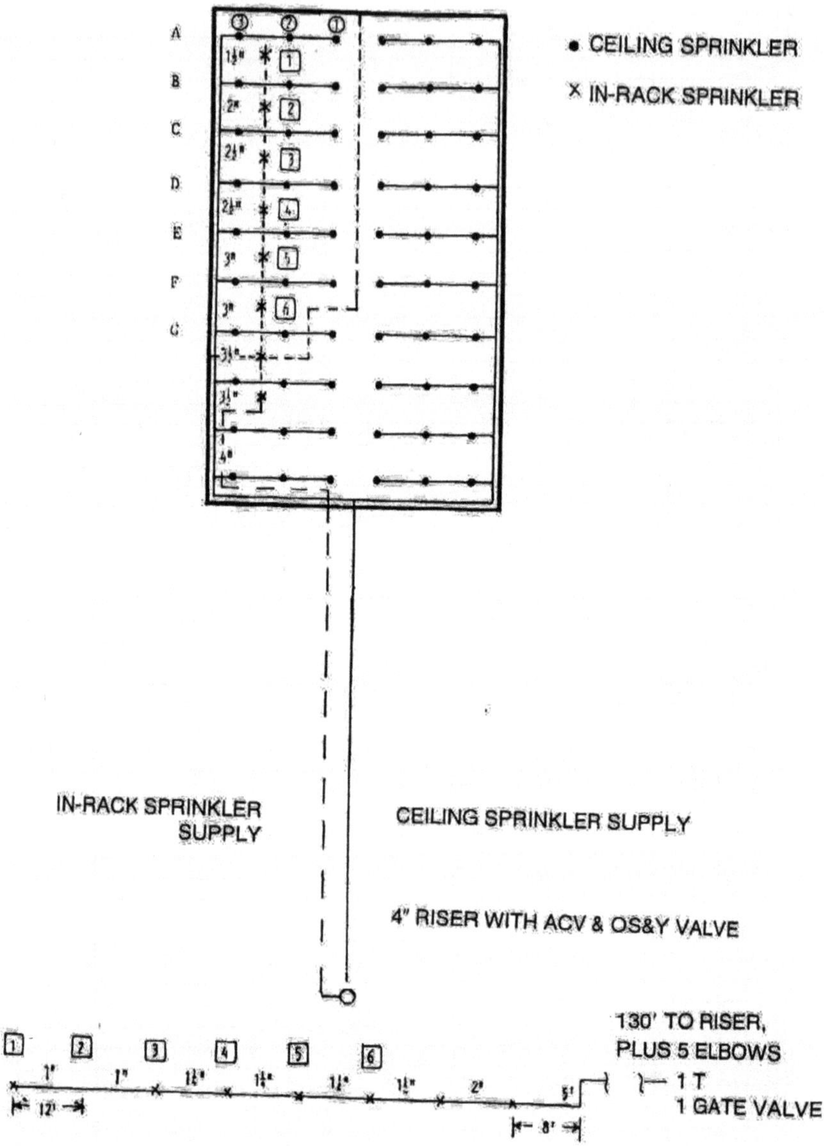

A
B
C
D
E
F
G
• CEILING SPRINKLER
× IN-RACK SPRINKLER
IN-RACK SPRINKLER SUPPLY
CEILING SPRINKLER SUPPLY
4" RISER WITH ACV & OS&Y VALVE
130' TO RISER,
PLUS 5 ELBOWS
1 T
1 GATE VALVE

# HYDRAULIC CALCULATIONS

NAME IN-RACK SPRINKLER CALCULATIONS
ADDRESS
AREA

SHEET ___ OF ___
LOC. NO.
DATE
BY
C = 120
K = 5.6

| Nozzle Ident. & Location | Flow in G.P.M | Pipe Size | Pipe Ftgs. & Devices | Equiv. Pipe Length | Friction Loss PSI/Ft. | Pressure Summary | Normal Pressure | SCHEDULE 40 PIPE Notes |
|---|---|---|---|---|---|---|---|---|
| 1 | q. 21.7 | 1" | | L. 12 | .151 | Pt. 15 | Pt. | |
| | | | | F. | | Pe. | Pv. | |
| | Q. 21.7 | | | T. 12 | | Pf. 1.81 | Pn. | |
| 2 | q. 21.7 | 1" | | L. 12 | .546 | Pt. 16.81 | Pt. 16.81 | |
| | | | | F. | | Pe. | Pv. 1.77 | AQ = 43.7 |
| | Q. 43.4 | | | T. 12 | | Pf. 6.55 | Pn. 15.04 | |
| 3 | q. 26.2 | 1¼" | | L. 12 | .344 | Pt. 23.36 | Pt. 23.36 | |
| | | | | F. | | Pe. | Pv. 1.50 | AQ = 69.4 |
| | Q. 69.6 | | | T. 12 | | Pf. 4.13 | Pn. 21.87 | |
| 4 | q. 27.7 | 1¼" | | L. 12 | .638 | Pt. 27.49 | Pt. 27.49 | |
| | | | | F. | | Pe. | Pv. 2.94 | AQ = 97.6 |
| | Q. 97.3 | | | T. 12 | | Pf. 7.66 | Pn. 24.54 | |
| 5 | q. 31.9 | 1½" | | L. 12 | .510 | Pt. 35.15 | Pt. 35.15 | |
| | | | | F. | | Pe. | Pv. 2.79 | AQ = 129.3 |
| | Q. 129.2 | | | T. 12 | | Pf. 6.12 | Pn. 32.36 | |
| 6 | q. 34.0 | 1½" | | L. 12 | .785 | Pt. 41.27 | Pt. 41.27 | |
| | | | | F. | | Pe. | Pv. 4.50 | AQ = 164.2 |
| | Q. 163.2 | | | T. 12 | | Pf. 9.42 | Pn. 36.77 | |
| | q. | 2" | | L. 12 | .233 | Pt. 50.69 | Pt. | Calculations ending at first head on in-rack line. |
| | | | | F. | | Pe. | Pv. | |
| | Q. 163.2 | | | T. 12 | | Pf. 2.80 | Pn. | |
| | q. | | | L. | | Pt. 53.49 | Pt. | |
| | | | | F. | | Pe. | Pv. | |
| | Q. | | | T. | | Pf. | Pn. | |
| | q. | | | L. | | Pt. | Pt. | |
| | | | | F. | | Pe. | Pv. | |
| | Q. | | | T. | | Pf. | Pn. | |
| | q. | | | L. | | Pt. | Pt. | |
| | | | | F. | | Pe. | Pv. | |
| | Q. | | | T. | | Pf. | Pn. | |
| | q. | | | L. | | Pt. | Pt. | |
| | | | | F. | | Pe. | Pv. | |
| | Q. | | | T. | | Pf. | Pn. | |
| | q. | | | L. | | Pt. | Pt. | |
| | | | | F. | | Pe. | Pv. | |
| | Q. | | | T. | | Pf. | Pn. | |

# A Bit of Ancient History—The Minimum Water Supply

Although it is now ancient history, we will mention the Minimum Water Supply because you may encounter an old system where this was included in the design. Until its demise in the 1980 Edition of NFPA 13, all Light and Ordinary Hazard systems were subject to a Minimum Water Supply introduced at the base of the riser. The required flow and pressure at the base of the riser was calculated in the normal manner. If the calculated flow at the base of the riser was less than the stipulated Minimum Water Supply, the flow at the base of the riser was increased to the Minimum Water Supply and this flow was carried through all upstream calculations. Following were the Minimum Water Supplies:

| | |
|---|---|
| Light hazard: | 150 gpm |
| Ordinary hazard group 1: | 400 gpm |
| Ordinary hazard group 2: | 600 gpm |
| Ordinary hazard group 3: | 750 gpm |

The Minimum Water Supply for ordinary hazard occupancies was equal to the maximum design area, 5000 sq. ft., times the specified density over this area, so the calculated flow was usually less than the Minimum Water Supply. (For the benefit of newcomers, the 1991 Edition of NFPA 13 combined Group 2 and Group 3 into a single Group 2, redrew the design curves and reduced the maximum Ordinary Hazard area to 4000 sq. ft.)

What was the logic behind the Minimum Water Supply? There was none, really. It apparently reflected a concern about the small flows for which some systems might be designed. In effect, it provided a small and sort of randomly varying cushion to the operating area. The cushion varied directly with the length of the sprinkler underground and, curiously, inversely according to the size of the sprinkler underground.

Concern for the size of the underground main supplying the system, in fact, could have entered into their thinking on the Minimum Water Supply. NFPA 13 requires that "the underground supply pipe shall be at least as large as the system riser" for pipe schedule systems, but a hydraulically designed system, rightly or



H. S. Wass Jr. and R. P. Fleming P.E., *Sprinkler Hydraulics*,
https://doi.org/10.1007/978-3-030-02595-3_28

wrongly, can have any size underground that can be supported by the hydraulic calculations. Thus, one reason behind the Minimum Water Supply may have been concern about very small underground mains.

The 1979 Edition of NFPA 231 for High-Piled Storage introduced a short-lived Minimum Water Supply for sprinklers (exclusive of the 500 gpm hose requirement) as follows:

| | |
|---|---|
| Class I and II commodity: | 400 gpm |
| Class III commodity: | 600 gpm |
| Class IV commodity: | 750 gpm |

This was subsequently dropped after NFPA 13 eliminated its Minimum Water Supply.

# Existing Sprinkler Systems—The Inspector's Problem: What Do We Have?

It is all very simple when you are in on the ground floor of a new calculated sprinkler system and the sprinkler contractor provides you with the sprinkler plans and the hydraulic calculations. But what happens down the road? Experience has shown that this is a major problem. Building owners and tenants typically do not appreciate the importance of retaining or obtaining a copy of the plans and calculations. Sprinkler contractors may only retain the plans and calculations for a certain number of years. Beyond that, sprinkler contractors go out of business. Other things happen. I have been told that the plans were lost in a flood.

NFPA 13 has not totally ignored this problem. From the beginning it has specified that basic design data should be displayed at the sprinkler riser and, more recently, it has required that hose stream demand also be stated. The standard currently specifies a "Hydraulic Design Information Sign" and provides the following sample wording in the Annex:

This system as shown on ....................................................... company print no .................................................................................. dated ............... for ...................................................................................... ................. at .................................................................................. contract no ................................ is designed to discharge at a rate of ...................................... $gpm/ft^2$ ........ ($L/min/m^2$) of floor area over a maximum area of ........................... $ft^2$ ($m^2$) when supplied with water at a rate of ..................... gpm (L/min) at .............................................. psi (bar) at the base of the riser. Hose stream allowance of ................................ gpm (L/min) is included in the above.

Occupancy classification ............................................... Commodity classification ........................................................ Maximum storage height.......................................................



H. S. Wass Jr. and R. P. Fleming P.E., *Sprinkler Hydraulics*,
https://doi.org/10.1007/978-3-030-02595-3_29

Unfortunately, while the text of the standard also requires that the "Hydraulic Design Information Sign" include "Location of the design area or areas", this sample wording does not clearly call for it.

In practice, the requirement for a hydraulic design placard or sign has been frequently ignored such that there is no information at the base of the riser. Beginning with the 2014 Edition of NFPA 25, however, a requirement was included that a missing or illegible hydraulic information sign must be replaced as part of any quarterly inspection, and that pipe schedule systems must also have signs identifying them as such. If you are lucky enough to find information at the base of the riser for a hydraulically designed system, then you know a little bit about it. However, you do not know if the calculations were properly performed. You *MUST* obtain reliable, preferably current, information on the available water supply. I have encountered a sprinkler system that was designed for a projected future water supply. Unfortunately, the anticipated water supply was never provided. As explained elsewhere in this book, if you have the design parameters (design density, design area, requisite flow, and pressure at the base of the riser and the elevation of the design area above the base of the riser), you can accurately estimate the density that can be delivered over the design area. Occasionally, you will need a different design area than that for which the system was calculated. I have been asked many times how you can handle this. The answer is the system must be recalculated, which means you must have a sprinkler plan.

With no design criteria and no plans, a calculated sprinkler system is a completely unknown quantity. If all else fails, it may be necessary to make up a sprinkler plan. Since a reasonable degree of accuracy is important, it is not acceptable to determine the distance between heads and branch lines by pacing and it is not acceptable to guess at the pipe size by "eyeballing." Where the sprinkler system is symmetrical, it may be possible to accurately determine the distance between heads and branch lines by measuring the interior of the building, or one bay of the building. A tape measure should be used unless building plans are available. Pipe sizes can be ascertained only by physically measuring the pipe. Since it is difficult to accurately measure the diameter without calipers, it may be best to measure the circumference of the pipe with a cloth tape. Of course, it is not always possible to get to the pipe, especially when there is a high ceiling. At this point, you may have to recommend that a sprinkler contractor be hired to do the job.

Refer to Table 3 in Appendix A for outside dimensions of pipe. As you can see in Table 3, however, knowing the outside diameter of the pipe may not tell you what kind of pipe you have. In the original edition of this book, published in 1983, we said: "if there are threaded connections, you can be sure it is Schedule 40 pipe." That is no longer true. In belated recognition of this problem, the 1994 Edition of NFPA 13 added the following:

> 2-3.7 Pipe Identification. All pipe, including specially listed pipe allowed by Section 2-3.5, shall be marked continuously along its length by the manufacturer in such a way as to properly identify the type of pipe.

This may not be foolproof, either. If they have just a single "continuous" line of identification, that line might end up on the top side of the pipe.

We have no quarrel with the changes of the past 50 years. Much has been learned about how to protect what we are putting in buildings. New kinds of pipe and joining methods offer economies both in terms of material cost and labor cost. This is all to the good. But the failure to seriously attempt to deal with the problem of knowing what has been installed at some future date has largely been ignored. Large corporations may have their own fire protection engineers who carefully maintain plans, calculations, and other details about their fire protection systems. Most large companies also are insured by major carriers who maintain this information. The problem is mainly with many small businesses. The management of these companies may have no awareness of the complexities of today's sprinkler systems. They may be comfortable with the knowledge that somewhere along the way some inspector gave the system his or her blessing. But not all Authorities Having Jurisdiction are created equal. The system may or may not have been properly designed. And, if it were, it may no longer be suited to the occupancy.

We are not prepared to offer a solution. We have a few thoughts but have not been able to refine them to a point where we can make a practical proposal. We can only suggest that all interested parties (or should we say parties who should be interested) should give this more attention than it has received up to now.

Before we move on, we will take note of one of the rare expressions of concern about the increasing complexity. The January 1993 issue of "Sprinkler Age", published by the American Fire Sprinkler Association (a trade association for sprinkler contractors) contained an article by its chairman, Don Becker.

> "A general maintenance employee may not recognize the difference between a quick response, extended throw sprinkler and a standard sprinkler. If the threads fit, they'll screw into the pipe whatever sprinkler is in hand.
>
> "As contractors, we are becoming victims of the manufacturers' options, and let's not leave out the recognized listing and approving organizations. We, as contractors, can in most cases properly apply the many different sprinkler head options provided to us, but what does the future hold for us as contractors when there is a loss of either life and/or property when an owner, years later, accidentally replaces sprinkler heads with some that were not in compliance with the application.
>
> "Isn't it time that the leaders of our industry from all sectors set some level of standardization and limits of selection...?"

We agree, except for his reference to "becoming victims of the manufacturers' options." A manufacturer responded, saying that "manufacturers' new products are often driven by the demands of the contractors." In any case, trying to assign blame serves no purpose. The present condition has simply evolved, driven by technological advances and economics. But it is time to look at where we are and ask if we really want to be here.

Progress has been made in recent decades, including the requirement within NFPA 13, effective January 1, 2001, that every unique model of sprinkler (indicated by a change in orifice, distribution characteristics, pressure rating or thermal sensitivity) be separately identified by means of 1 or 2 alphabetic characters representing the manufacturer, and 3 or 4 numbers. The identification of the sprinkler manufacturer codes is available at www.ifsa.global. Individual manufacturers and product certification agencies are expected to be able to identify the characteristics associated with a particular numerical suffix. These SIN (Sprinkler Identification Number) codes can also be used to verify that the sprinklers shown on the drawings are the same as those installed in the field.

# Hose Streams

Hose streams have received only passing reference until now, but they are an important element in a calculated sprinkler system. The relationship between the output of a sprinkler system, that is, the density over a design area, and the water supply has been explained. A realistic evaluation must also recognize the anticipated "people" response to a fire and superimpose this upon the "automatic" sprinkler response. If there are inside hose stations, they may be in use at the same time the sprinkler system is doing its work. Fire department response is hoped for and must be assumed. Hose streams brought into play by building occupants or the fire department deplete the water supply, which is the critical parameter of all calculated sprinkler system designs.

All sprinkler design standards include a hose-stream allowance intended to compensate for what is judged to be reasonable anticipated hose-stream use. The fact that "reasonable anticipated …use" is *just that* demonstrates once again that this whole "scientific" process is fraught with unknowns and uncertainties. Further, recognize that hose-stream demands in the standards are *minimum,* and it may be entirely appropriate to use judgment and specify a higher hose-stream demand. NFPA 13 specifies a "combined inside and outside hose" allowance for Light Hazard, Ordinary Hazard, and Extra Hazard systems designed to their criteria. Former standards NFPA 231 and 231C both stipulated an allowance of 500 gpm for hose streams for highpiled storage. The "outside" hose allowance contemplates fire department usage at nearby hydrants, while the "inside" or "small hose" is the 1½-in. hose supplied through the sprinkler piping for manual fire-fighting within the occupancy.

The 1980 Edition of NFPA 13 addressed, for the first time, how much to allow for inside hose and subsequent editions refined the method of incorporating the hose stream(s) into the calculations.

5.2.3.1.3(d) When inside hose stations are planned or are required by other standards, a total water allowance of 50 gpm for a single hose station installation or 100 gpm for a multiple hose station installation shall be added to the sprinkler requirements. The water allowance shall be



H. S. Wass Jr. and R. P. Fleming P.E., *Sprinkler Hydraulics*,
https://doi.org/10.1007/978-3-030-02595-3_30

added in 50 gpm increments beginning at the most remote hose station, with each increment added at the pressure required by the sprinkler system design at that point.

With respect to outside hose requirements, NFPA 13 stated:

5.2.3.1.3(f) Water allowance for outside hose shall be added to the sprinkler and inside hose requirement at the connection to the city water main, or at a yard hydrant, whichever is closer to the system riser.

We agree, with one exception. Following the "real life" approach, if the city supply is a dead-end main and the only hydrant that can reasonably be expected to be used is on the supply side of the connection between the sprinkler underground and the city main, it is only logical to go back to that hydrant on the city main before adding in the allowance.

Untill now, our discussion of inside hose has been confined to 1½-in. hose that may be used by the building's occupants. Now we will consider 2½-in. hose outlets intended for fire department use. Traditionally, the 2½-in. hose outlets have been associated with standpipe systems that were independent of sprinkler systems and usually found in high-rise buildings. It is permissible, however, to attach 2½-in. hose outlets to "combined risers" that serve both the standpipe outlets and sprinkler system piping.

Standpipe systems are addressed by NFPA 14 - Standard for the *Installation of Standpipe and Hose Systems.* NFPA 14 systems have heavy water supply requirements (normally 100 psi outlet pressure and a volume ranging from 500 to 1000 gpm). In a fully sprinklered building, the NFPA 13 water demand (sprinkler and hose) is not additive to the NFPA 14 demand. The total water demand is the higher of the two, which is usually the NFPA 14 demand. In a partially sprinklered building it is specified that the sprinkler demand (but not the associated hose demand) should be added to the NFPA 14 water demand. The NFPA *Automatic Sprinkler Systems Handbook* explains that "this is necessary since there is no way to prevent fires from originating in the nonsprinklered area." Presumably, the thought is that a fire in the unsprinklered area might subsequently spread a fire into, or open heads in, the sprinklered area, incurring an additional demand upon the water supply.

The guidelines for incorporating the various kinds of hose streams into the sprinkler system calculations are, of necessity, very general. Good judgment applied to a specific site might suggest deviations from the codes. While the minimum code requirements should normally be complied with, there is certainly no reason why additional water demand could not be incorporated if conditions warrant.

Before leaving this subject, return to the example appearing on page 103. We estimated an actual available density of about 0.22 gpm per sq. ft., without making allowance for hose streams. Now assume that 500 gpm should be reserved for hose streams. On page 180 is the same graph that is on page 103. The water-supply "curve," when plotted on the conventional 1.85 paper, is normally a straight line.

When an allowance of 500 gpm is—deducted from the water supply, the resultant plot becomes a curve. You can determine this for yourself by arbitrarily selecting a number of points along the water-supply curve, deducting 500 gpm (or whatever allowance you wish to make), and making a dot for the reduced gpm *at the same pressure.* Since we are only interested in the effect of hose streams on the sprinkler discharge, we want to locate the point at which the curve representing water supply less hose streams crosses the characteristic curve for the sprinkler discharge area. To do this, plot a couple of points in the vicinity of the characteristic curve and connect them with a straight line (if the line is short, a straight line is a reasonable approximation of a curve). In the example on page, the intersection point is at a flow of about 465 gpm, using the same method used on page 103,

$$\frac{465}{592} \times 0.25 = 0.20\, gpm\, per\, sq.\, ft.$$

It might be noted that the water supply in our example is a fairly “flat” curve. When the water supply is weak, with a steep curve, the effect of hose streams on the sprinkler discharge is more pronounced.

See the illustration on page 181.

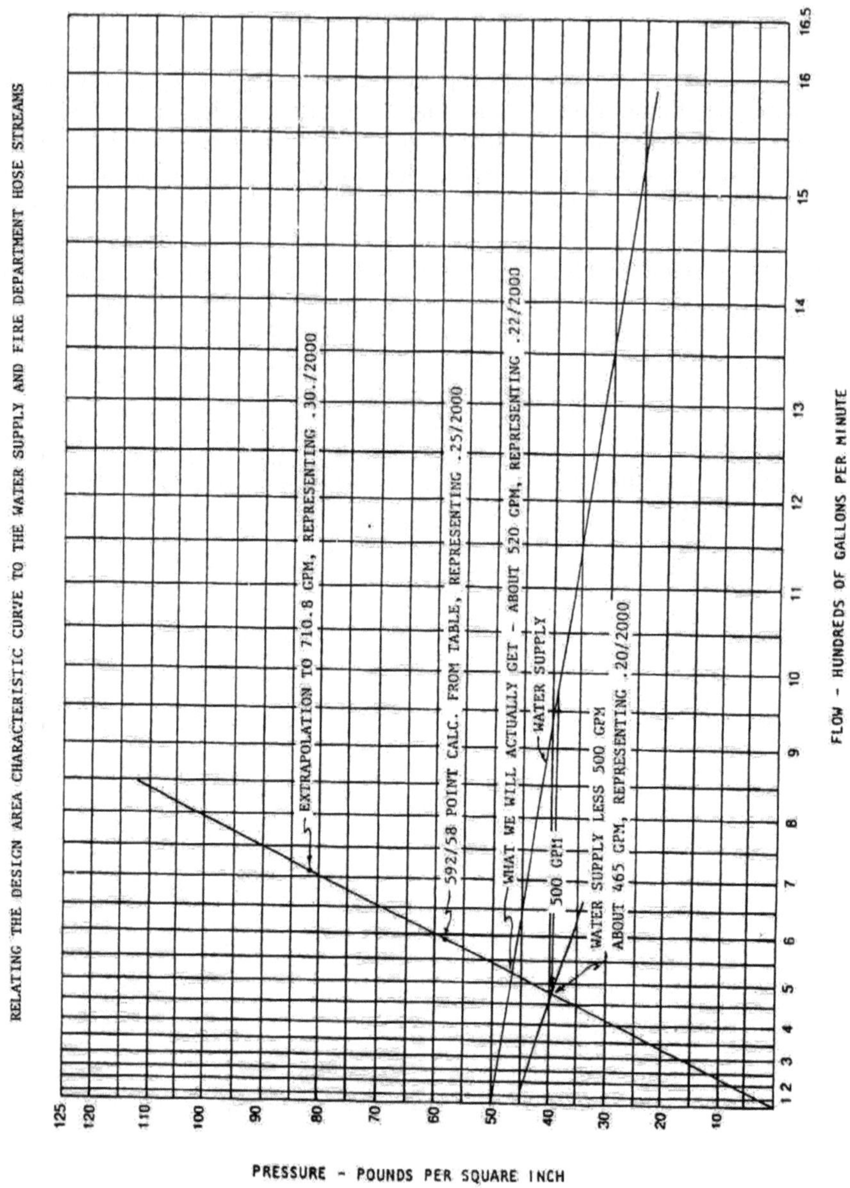
RELATING THE DESIGN AREA CHARACTERISTIC CURVE TO THE WATER SUPPLY AND FIRE DEPARTMENT HOSE STREAMS
EXTRAPOLATION TO 710.8 GPM, REPRESENTING .30./2000
592/58 POINT CALC. FROM TABLE, REPRESENTING .25/2000
WHAT WE WILL ACTUALLY GET - ABOUT 520 GPM, REPRESENTING .22/2000
WATER SUPPLY
500 GPM
WATER SUPPLY LESS 500 GPM
ABOUT 465 GPM, REPRESENTING .20/2000
FLOW - HUNDREDS OF GALLONS PER MINUTE
1 2 3 4 5 6 7 8 9 10 11 12 13 14 15 16 16.5
PRESSURE - POUNDS PER SQUARE INCH
125 120 110 100 90 80 70 60 50 40 30 20 10

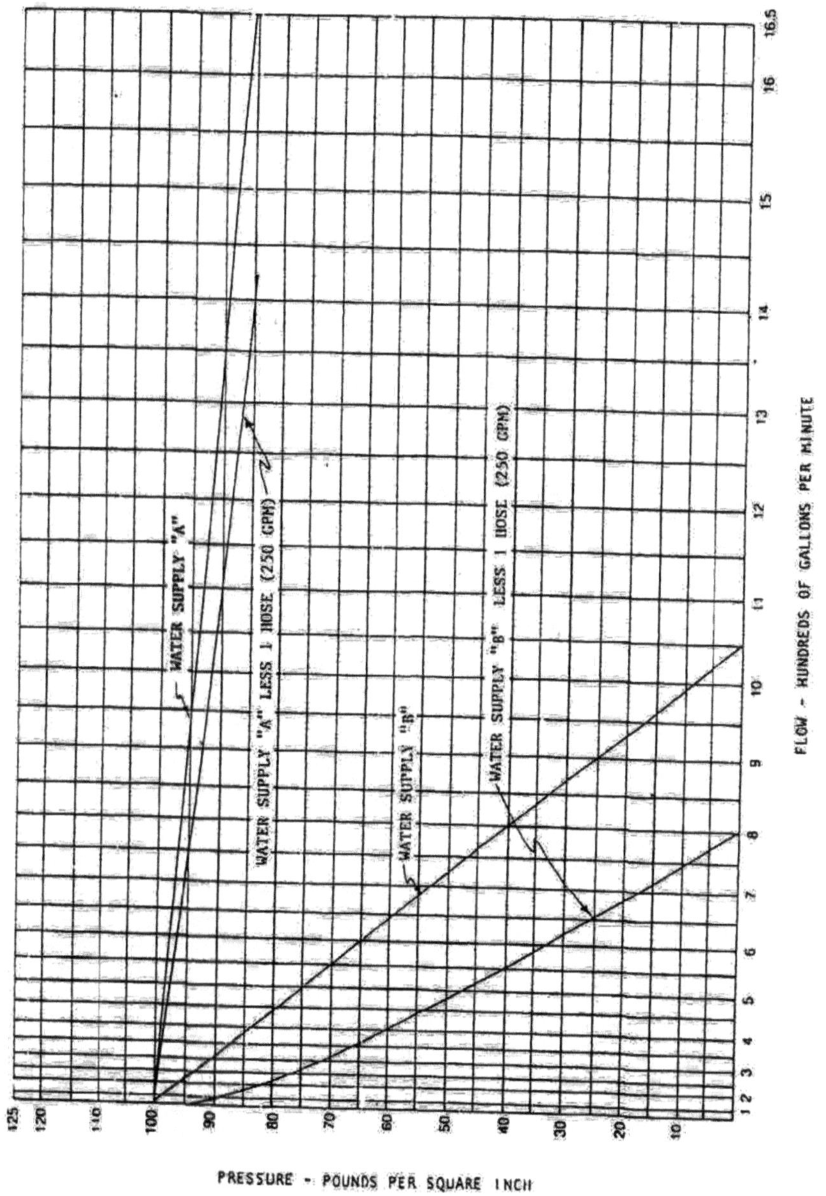


ILLUSTRATION OF THE EFFECT OF A HOSE STEAM ALLOWANCE ON DIFFERENT WATER SUPPLIES

# The Water Supply Problem

It should be evident by now that, however, well a system is designed, if the necessary water supply is not available when needed, all of the fine work has been in vain. At the outset, we suggested that the water supply was capable of compromising all of our fancy figures. We will elaborate.

Only very limited credence should be given to estimates of the water supply based upon two-inch drain tests, although formulas have been devised, published and sometimes used for this purpose. A drain test gives an experienced person a general idea of the water supply but it should not be used as a basis for hydraulic calculations. Again, a single hydrant flow test is sometimes used to determine a water supply but only a properly conducted hydrant flow test, as discussed previously, provides a sound basis for calculations. Even a properly conducted flow test (with, of course, accurate gauges) only provides an indication of the water supply at the moment the test is performed. Always be aware of the following:

1. Public water supplies can vary from hour to hour or from day to day because pumps may be on or pumps may be off or gravity supplies may be turned on or turned off.
2. The water supply is affected by local use, which is likely to vary with the time of day and with the season of the year. Occasionally, you will encounter a water department that will only allow tests to be made at night. A test at night may not be indicative of the supply during the day when, for example, there may be higher water use by the local industry. A test in the spring or fall may not reflect the supply during a dry spell in the summer when many people are watering their lawns.
3. In a rapidly growing area, the demand and, consequently, the available water supply can change significantly in a few years.
4. Changes may be made in the water system. While these changes usually will result in an improvement in the water supply, there are instances where the pressure is reduced even if the available volume is increased.



H. S. Wass Jr. and R. P. Fleming P.E., *Sprinkler Hydraulics*,
https://doi.org/10.1007/978-3-030-02595-3_31

5. The so-called static pressure is really not a true static pressure. In a public water main, it is actually a residual pressure with an unknown flow. As a consequence, when the hydrant flow test is plotted in a conventional manner, the real water supply probably falls slightly below the line at lower flows and slightly above the line when it is extrapolated above the test flow. See the simplified example on page 185.
6. Many possible inaccuracies are inherent in the hydrant flow test procedure. The quality of the gauges varies. How have they been treated by the user? How recently have they been calibrated? Has the pitot reading been taken at the proper location (at the center of the stream and a distance equal to about one-half the diameter of the outlet)?

It is reasonable to conclude that the preceding considerations should be recognized when designing a sprinkler system. Either "good engineering judgment" should be applied or ignorance of the possible variables should be acknowledged with a safety factor added to the tested water supply.

NFPA 291, Fire Flow Testing and Marking of Hydrants, has a discussion of considerations relative to the water distribution system, which is recommended reading, but is silent on the problems we are discussing here. NFPA 13 also fails to address the subject. As a result, the sprinkler contractor normally designs the sprinkler system to take full advantage of the water supply that was indicated by the test. Arbitrary safety factors, such as 5 psi below the supply curve, are sometimes required by Authorities Having Jurisdiction, but there is a need for a well-thought-out and generally accepted approach to the problem. We are not prepared to offer one at this time but wiser heads should take a break from their pursuit of ever-greater theoretical understanding long enough to address one of the real-world weaknesses in the application of the theory.

There is another matter worth mentioning. When computing the flow from the common smooth and rounded hydrant outlet, it is general practice to use the discharge coefficient of 0.90. When the hydrant outlet is square and sharp, a 0.80 coefficient is generally used, and when it is square and projecting into the barrel, the customary coefficient is 0.70.

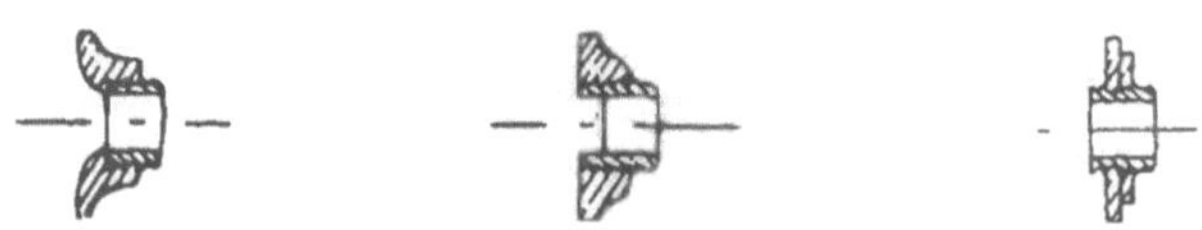

Smooth and rounded    Square and sharp    Projecting into barrel

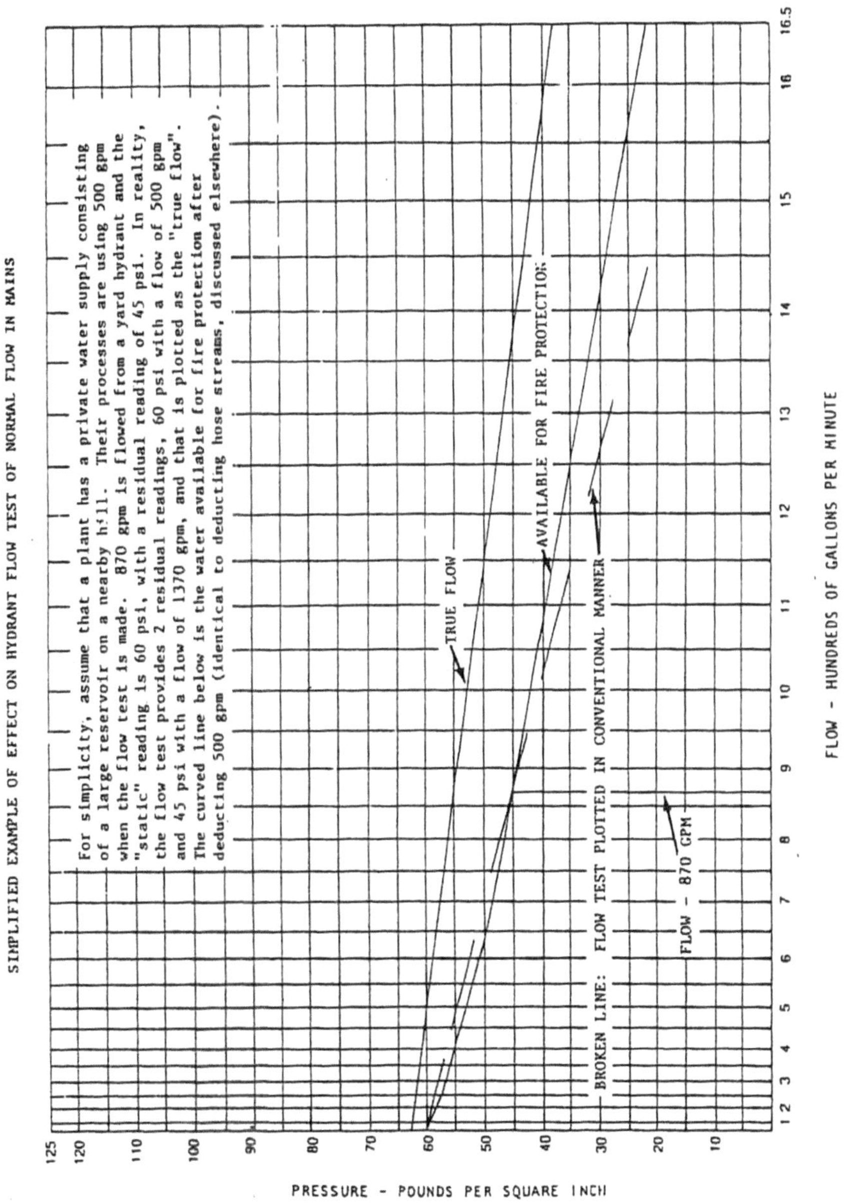
SIMPLIFIED EXAMPLE OF EFFECT ON HYDRANT FLOW TEST OF NORMAL FLOW IN MAINS
For simplicity, assume that a plant has a private water supply consisting of a large reservoir on a nearby hill. Their processes are using 500 gpm when the flow test is made. 870 gpm is flowed from a yard hydrant and the "static" reading is 60 psi, with a residual reading of 45 psi. In reality, the flow test provides 2 residual readings, 60 psi with a flow of 500 gpm and 45 psi with a flow of 1370 gpm, and that is plotted as the "true flow". The curved line below is the water available for fire protection after deducting 500 gpm (identical to deducting hose streams, discussed elsewhere).
TRUE FLOW
AVAILABLE FOR FIRE PROTECTION
BROKEN LINE: FLOW TEST PLOTTED IN CONVENTIONAL MANNER
FLOW - 870 GPM
125 120 110 100 90 80 70 60 50 40 30 20 10
PRESSURE - POUNDS PER SQUARE INCH
1 2 3 4 5 6 7 8 9 10 11 12 13 14 15 16 16.5
FLOW - HUNDREDS OF GALLONS PER MINUTE

All of this is affirmed in NFPA 291. We have been told, however, that tests made on different makes of hydrants have indicated various coefficients for the common "0.9" hydrant ranging to below 0.80. Therefore, in the absence of good data on the appropriate coefficient for each make and model (we wonder why the data are not available), an argument can be made for using a more conservative set of coefficients, perhaps 0.80, 0.70, and 0.60, respectively. Better yet, perhaps, discharge through play pipes with 1–3/4 in. nozzles where a coefficient of 0.97 can be applied with some degree of confidence.

# Reliability of Automatic Sprinkler Systems

In our statistic-ridden society, it is only natural to encounter statistics on the performance of sprinkler systems. We will spare you a philosophical discussion of how facts are frequently used to obscure the truth. But speaking as one who has been trapped into mouthing reliability statistics on a major New York television station, we hope you will believe us when we say that all such numbers have little meaning. So we will not quote from the NFPA Handbook or a study made in Australia.

Do not try to quantify, but be fully aware of reliability considerations. What are the reliability considerations?

First, it is necessary to define what we mean by reliability. In its broadest sense, reliability can be defined as the probability that the sprinkler system will perform successfully in the event of a fire. Right away there is a problem because it is difficult to define what is meant by "successful" performance. In a general way, with the exception of an ESFR system, successful performance can be defined as control, not necessarily extinguishment, of a fire. In some instances, a single sprinkler will extinguish a fire. In some high fire loading situations, or where shielded conditions exist, 20, 30, or more sprinklers could operate and final extinguishment would have to be accomplished by the fire department, yet it could still be considered a "successful" performance.

Sometimes the sprinkler system does not control a fire; that is, the sprinkler performance is deemed unsuccessful. Why does this happen?

At what might be called the first level of reliability is the question of whether or not the system meets the generally accepted standards—NFPA 13 or other appropriate standard. This first level involves the following:

1. Full sprinkler coverage, as outlined in NFPA 13; that is, sprinklers in all areas specified therein, in accordance with all of their rules pertaining to spacing, concealed spaces, shielded areas, etc.
2. Adequate anticipated (based upon an appropriate current flow test) water supply.
3. Adequate sprinkler design, in terms of water supply and occupancy.



H. S. Wass Jr. and R. P. Fleming P.E., *Sprinkler Hydraulics*,
https://doi.org/10.1007/978-3-030-02595-3_32

After evaluating the first level of reliability, what are the considerations at the second level of reliability?

1. Possible conditions not contemplated in the design. A few examples are:
   a. Multiple fires set by an arsonist.
   b. A fire involving a transient flammable liquid not normally found on the premises.
   c. An interior finish with special characteristics that have not been recognized. An example that surfaced some years back was some forms of aluminum-foil-paper sandwich insulation in common use in California in which a rapid combustion process could be initiated.
2. An unknown impairment of the sprinkler system; most commonly, a closed or partially closed valve. Another possible problem would be foreign matter in the piping that is carried to the discharging head, obstructing the flow.
3. A known temporary impairment of the sprinkler system. Common examples are alterations to the system, mechanical damage inflicted by a forklift, a separation of an improperly installed fitting, a freeze-up due to the failure to maintain heat in a wet system or failure to drain low-point drains in a dry system.
4. Impairment of the public water supply, where, again, there could be a closed or partially closed valve or, perhaps, several valves. The effect of a single closed valve on the public supply depends upon the arrangement of the supply. A single closed valve on a dead-end main can mean total loss of supply; on a grid, a reduced supply. Aside from the closed valve, it is necessary to consider the source or sources of the water. Even a large municipal pumping station, which seemingly has adequate redundancy, may not be immune to total failure, perhaps due to flooding.
5. Impairment of the private water supply. Private water supplies exist for one of three reasons:
   a. There is no public water supply available.
   b. The public water supply is inadequate.
   c. The public water supply is adequate, but the redundancy of a private supply has been provided to enhance reliability.

Aside from the ubiquitous closed-valve problem, a private water supply can be impaired in other ways, such as:

a. Freezing, due to failure of the heating system.
b. Failure of a pump to start. Electric-driven pumps are fairly reliable, but will not be available if there is a power failure (in the absence of a back-up generator). Diesel-driven pumps are subject to more problems that could prevent them from operating, but are not dependent upon an external source of power.

Although somewhat obvious, the following are the main approaches to increasing reliability:

1. Testing and inspection programs. NFPA 25 provides the guidelines for adequate inspections and tests.
2. Security to prevent deliberate compromising of the system.
3. Strictly enforced procedures for necessary system shutdowns.
4. Supervision; that is, automatic monitoring of such things as valve status, tank temperature, and water levels.
5. Redundancy.

No discussion of reliability should ignore the booster pump. Much is heard about booster pumps, particularly when the existing protection is little weak. Perhaps, a booster pump will solve the problem. We suggest that the very word "booster", with all of its positive connotations, has a good sound to it. And of course, a booster pump will push more water through the sprinklers.

While we will get to the reliability connection, a brief digression:

1. A booster pump does not increase the volume of water available, only the pressure. As a general rule, it is not desirable to decrease the pressure in a water main below 20 psi. Assuming 20 psi minimum allowable pressure in the main, when a hydrant flow test is plotted on the 1.85 semi-exponential graph paper and the line is extended to the point where it crosses 20 psi, that is the volume of water available, with or without a booster pump. Recognize, however, that we are talking about the volume available at the point where the booster pump is located. Therefore, the farther upstream, that is, toward the supply, that the pump is located, the greater the volume available for the output of the pump and to that extent, it could increase the volume available at the point where it is needed.
2. Higher pressure at the sprinkler means higher discharge from the sprinkler and, consequently, a higher density. As mentioned previously, however, higher pressure means smaller droplets and at some point, there may be a problem with the fine droplets penetrating the updraft from a fire. We probably should be wary of a comfortable sprinkler density achieved by a pressure above 100 psi.
3. Finally, we reach the reliability problem. A booster pump may not operate when it is needed, for many reasons. Precise probabilities cannot be assigned, but consider the wide range of things that could go wrong and make your own judgment.

The booster pump belongs in the repertoire of fire-protection hardware, but we suggest that it should be used very judiciously. As an example, we once received a call from a sprinkler contractor asking if we would accept a booster pump on the water supply to a large shopping center. He stated that he could more than make up the cost of the pump by the reduced size of the sprinkler piping with a hydraulic design utilizing the higher pressure. We advised him that a booster pump was not acceptable. If it is practical to design a system utilizing existing pressure, reliability should not be sacrificed by the addition of a booster pump.

Sometimes a booster pump is necessary or, on balance, desirable. When might a booster pump be "desirable"? One example would be when the high flow and pressure requirements for an ESFR system in a warehouse exceed the capability of an otherwise good water supply. While the water supply would be adequate for a conventional system, the tolerance of an ESFR system to changes in storage height and commodity types plus the probability that a fire would result in a smaller loss might tip the scales toward the booster pump. When we use a booster pump, redundancy should always be considered. Typically, redundancy consists of one electric and one diesel pump installed in parallel, with the diesel pump designed to kick in if the electric pump fails to start If you are protecting the local supermarket you will probably take your chances with a single pump. If you are protecting a 750,000 sq. ft. regional warehouse supplying the supermarket you should insist on some level of redundancy. There are no hard and fast rules but you must judge the level of reliability against the consequences of an unsuccessful sprinkler performance.

# The Use and Abuse of the "K"

It is common practice, when calculating more than one identical branch line, to calculate the end branch line, then compute a "*k*" for the entire branch line. Each remaining branch line is treated as if it were a single big sprinkler and the branch line flow calculated using the computed "*k*".

To illustrate how this is done, suppose that when the end branch line is calculated, you arrive at a flow of 140 gpm at 36 psi at the point where the branch line connects to the cross main.

$$Since\, Q = k\sqrt{p},\; k = \frac{Q}{\sqrt{P}} = \frac{140}{\sqrt{36}} = 23.333$$

The pressure at each other branch line is used in conjunction with this "*k*" to compute the branch-line flows.

It should be recognized that this method is not strictly correct. The flow in the branch line varies according to the square root of the pressure, but friction loss varies according to the flow to the 1.85 power. To convert this to the relationship between friction loss and pressure, use the reciprocal of 1.85, which is approximately 0.54. Thus, the flow through a branch line involves both the relationship $Q = kp^{0.5}$ and $Q = kp^{0.54}$. Therefore, the correct exponent for "*p*" when considering a flowing branch line is somewhere between 0.5 and 0.54. With a large pipe and short branch lines, the value would be closer to 0.5. With high friction loss resulting from a small pipe, it would tend toward 0.54. Also, holding the pipe size and length constant, the higher the flow, which means higher pressure, the more the true value would tend toward 0.54. Short branch lines, however, are likely to be closer to 0.50 even with high flow.

FM's rule of thumb,(Factory Mutual, now FM Global), called for the use of 0.54 when the density was 0.20 or more, and yielded conservative results. There is no need for this refinement when you are making relatively small adjustments to flow and pressure, and we are inclined to feel that the simple relationship,



H. S. Wass Jr. and R. P. Fleming P.E., *Sprinkler Hydraulics*,
https://doi.org/10.1007/978-3-030-02595-3_33

$$\frac{Q_1}{Q_2} = \sqrt{\frac{P_1}{P_2}}$$

is satisfactory for most purposes.

We question if *k*'s for branch lines should be used for final calculations by a computer. Further, perhaps they should not be used unnecessarily in hand calculations. We have seen "*k*" used in hand calculations for a one-sprinkler branch line where the proper trial-and-error process is quick and easy. On the other hand, comparisons we have made suggest that the error resulting from the "*k*" is not likely to be significant.

There is one other thing to remember about "*k*"s. They are technically not valid where there are elevation differences between the discharging sprinklers, since there is an extraneous constant (elevation) affecting the "*p*" and the "*p*" is being manipulated as a variable. Again, if the elevation differences are not great, the error is probably not significant.

In short, the "*k*" can be a useful tool in making calculations and, in many cases, will provide a reasonable approximation, but it should not be used indiscriminately.

# What Does It All Mean?

The preceding pages contain a lot of equations and a lot of numbers. Nicely calculated numbers can present two dangers. They can create a facade of knowledge and precision when, in fact, more subtle judgments are needed. We have suggested just that at various points along the way but we want to say it again. On the other hand, those who are easily intimidated by equations, numbers, and theory can be motivated, probably subconsciously, to emphasize the uncertainties and argue with great sincerity that we should simply "get back to basics."

The landscape is dotted with dramatic failures of sprinkler systems.

Unfortunately, the ensuing stories about these fires are not always accurate. The Sherwin-Williams fire in Dayton, Ohio some few years back comes to mind. This was a fully sprinklered flammable and combustible liquids warehouse. The sprinklers were totally ineffective. A graphic videotape was produced which was well worth watching. Some otherwise intelligent discussion of the fire was marred when an individual being interviewed said something to the effect that the sprinkler systems were designed in accordance with NFPA codes. This was not true. In fact, the relevant code, NFPA 30, did not address the kind of high-piled storage that was involved because nobody knew how to protect it. Such careless comments can unfairly undermine the credibility of the NFPA codes and standards.

Subtle judgments, referred to earlier, must come into play when evaluating sprinkler protection that does not fully comply with the applicable NFPA codes or other reputable protection guidelines. Not all deficiencies are as flagrant as the one just cited.

Attempts have been made over the years to quantify some kind of "efficiency rating" for sprinkler systems, to quote a term ISO (Insurance Services Office) coined some years ago. Without delving into the mechanics, the methodology, of course, involved manipulating numbers relating to the water supply, the sprinkler design, and sprinkler demand. The end result might be, for example, that the sprinkler protection met 86% of the desired level of protection. Assuming that the water supply and sprinkler design information is accurate and the sprinkler demand numbers correctly reflect the current state of knowledge, what does such a number mean?



H. S. Wass Jr. and R. P. Fleming P.E., *Sprinkler Hydraulics*,
https://doi.org/10.1007/978-3-030-02595-3_34

It is hard to say. Clearly, an "efficiency rating", or whatever you choose to call it, of 86% is better than a rating of 70%. That is, a higher level of confidence can be attached to the one over the other. How much confidence?

Well, perhaps that is why insurance companies have underwriters. When something cannot be quantified in a clear meaningful way by the engineer, the engineer must yield to someone who is comfortable working in ignorance.

What are we trying to say? Try to understand the current technology, including the underlying elements such as sprinkler hydraulics. Recognize the benefits that can be derived from the progress that has been made. Also, recognize that the current technology has its limits. It provides a starting point, a point of departure, when evaluating a specific fire protection problem. While it does not provide all of the answers, the failure to utilize the knowledge that has been accumulated is responsible for most of the sprinkler system failures. That may be our greatest challenge.

# A Little Learning

*A little learning is a dang'rous thing Drink deep or taste not the Pierian spring. There shallow draughts intoxicate the brain And drinking largely sobers us again.*

—Alexander Pope *An Essay on Criticism*

Understanding the theory and mechanics set forth in this book provides only a "little learning." Density, the heart of most calculated systems, is, perhaps, deceptively simple. Since density relates to the amount of water being discharged by the sprinkler system, it obviously is critical to the ability of the sprinkler system to control a fire. But, as we discussed elsewhere, what is actually being delivered to the seat of the fire is much more critical than what is being discharged. The pertinent properties of water are quite simple. The heat of vaporization, that is, the amount of heat required to convert water to steam is 970 BTU per pound. (BTU stands for British Thermal Unit and is the amount of heat required to raise the temperature of one pound of water one degree F). Therefore, a pound of water at a temperature of 55 degrees F introduced into a fire will absorb 157 BTU (212−55) to raise it to the boiling point and an additional 970 BTU to convert it to steam, a total of 1127 BTU per pound of water. A gallon of water weighs about 8.34 lb, so it requires about 9400 BTU to convert one gallon of water to steam. The BTU content of a pound of ordinary combustibles, wood or wood-derived products, varies, but is usually slightly less. In a general way, one gallon of water will absorb the heat from the combustion of one pound of wood. Of course, all water reaching the fire won't necessarily be converted to steam. It might be noted that the exothermicity (to throw in a pretentious word you may encounter in the literature) of plastics is about twice that of wood products.

If you are familiar with large-drop and ESFR sprinkler systems (now known as CMSA and ESFR sprinkler systems) you know that density is never mentioned in their design. It is the discharge pressure that matters because we are concerned about droplet size and droplet momentum. While density is not mentioned, it is still lurking in the background. With a "Maximum Protection Area" of 100 sq. ft. per sprinkler, a k-14 ESFR system with a minimum discharge pressure of 50 psi has a density of about 1.00 gpm per sq. ft. Density is still there, even if they choose not to mention it.



H. S. Wass Jr. and R. P. Fleming P.E., *Sprinkler Hydraulics*,
https://doi.org/10.1007/978-3-030-02595-3_35

The hydraulic calculations tell you how much water will come out of the sprinklers, but they do not tell you when the heads will actuate or what happens after that. Consider a few of the variables:

1. Sprinklers traditionally have had a dual role in a fire. They must be able to deliver water to the burning combustibles and to the combustibles adjacent to the fire to initiate the control/extinguishing process. At the same time, they must deal with the effects of the fire; that is, the liberated heat along the underside of the ceiling or roof. Failure to deal with this heat could mean excessive heads opening away from the fire, ignition of the ceiling or roof if it is combustible, or structural failure if there is exposed steel. The two roles of sprinklers are not entirely compatible. The heat at the ceiling can be most effectively absorbed by fine droplets, but fine droplets, performing that function may not survive to get to the seat of the fire. Also, any surviving fine droplets may not be able to penetrate the updraft from the fire. As we have said, increasing pressure will increase the density and decrease the droplet size. Higher pressure is not necessarily desirable and, certainly, a discharge pressure above 100 psi should be viewed warily.
2. The vertical distance between the sprinkler and the fire is an important consideration. Greater distance means greater delay in the sprinklers operating and less water getting to the fire after they do operate. Also, with a low ceiling, the sprinkler discharge will return some of the products of combustion to the fire, lowering the oxygen content. This is recognized in the credit given for quick response sprinklers. The reduction in the size of the sprinkler operating area is based on ceiling height.
3. Sprinklers do not always open where they are needed in the immediate area of the fire. With an intense fire and a strong updraft (perhaps 30–40 mph or stronger), the discharge from the first few sprinklers that open may be driven along the ceiling and keep adjacent sprinklers cool while sprinklers well removed from the fire area start opening. This phenomenon, known as "skipping," can prevent the nicely calculated density from being achieved in the critical immediate fire area. This is seen as an advantage for new electronically-activated sprinklers, in which a grouping of sprinklers can be actuated by a control panel based upon fire detection provided other than through the sprinklers' thermal elements.
4. Air currents are important. The size and tightness of the building affect the incoming supply of fresh oxygen. (The burning of one pound of wood, alluded to at the start of this section, might need the oxygen from a bit over 8 cubic feet of air.) Air-handling systems can move the products of combustion, opening sprinkler heads away from the fire. Automatic roof vents, and perhaps associated draft curtains, can have all sorts of consequences, perhaps favorable, perhaps unfavorable. (Roof vents in sprinklered buildings have been inconclusively studied and debated for many years. The only comment we will make is that we do not favor automatic vents in a sprinklered building.)

5. The temperature rating of the sprinklers has a bearing on the sprinklers that open in a fire. So-called high-temperature (286 °F) sprinklers are favorable in many high fire loading situations relying on traditional fire control by sprinklers. The time lag between the opening of standard 165 °F sprinklers and 286 °F sprinklers may be negligible because of the rapid heat buildup while, at the same time, fewer sprinklers away from the fire area may open when they are high-temperature rated. On the other hand, with a less rapidly developing fire, there may be a significant delay with high-temperature sprinklers and better results may be achieved with the standard 165 °F sprinklers. It is not clear where the dividing point is, but be wary of high-temperature sprinklers when the specified density is below 0.25 gpm/sq.ft. (reflecting a lesser challenge). Newer suppression-oriented sprinklers like ESFR do not rely on fire control, meaning that high temperature ratings are not needed to prevent excessive sprinkler activations.
6. The response time of the sprinklers is significant. Traditional sprinklers have response times that vary considerably. Tests have shown that when quick response sprinklers are used fewer sprinklers normally open. However, NFPA 13 still prohibits their use in Extra Hazard and other occupancies where there are substantial amounts of flammable liquids or combustible dusts out of concern for excessive sprinkler activations.

The foregoing is not a comprehensive treatment of the many variables affecting actual sprinkler operation, but simply provides a glimpse of the complexities. Reasonably accurate hydraulic calculations are important, but this is only one of many elements in designing or evaluating sprinkler protection.

# Appendix A

Specific friction loss values or equivalent pipe lengths for alarm valves, dry-pipe valves, deluge values, strainers and other devices should be obtained from the manufacturer. Also, specific friction loss values or equivalent pipe lengths for listed fittings should be used where they differ from the above table. This can be important. Some specially listed saddle type fittings, for example, have a friction loss considerably greater than shown in the above table.

**Table A.1** Equivalent pipe length chart for valves and fittings

NFPA No. 13 offers the following guidance: Equivalent pipe length chart (EPL) for Schedule 40 steel pipe (Feet)

| Pipe size (in.) | ¾ | 1 | 1¼ | 1½ | 2 | 2½ | 3 | 3½ | 4 | 5 | 6 | 8 | 10 | 12 |
|---|---|---|---|---|---|---|---|---|---|---|---|---|---|---|
| 45° elbow | 1 | 1 | 1 | 2 | 2 | 3 | 3 | 3 | 4 | 5 | 7 | 9 | 11 | 13 |
| 90° standard elbow | 2 | 2 | 3 | 4 | 5 | 6 | 7 | 8 | 10 | 12 | 14 | 18 | 22 | 27 |
| 90° long turn elbow | 1 | 2 | 2 | 2 | 3 | 4 | 5 | 5 | 6 | 8 | 9 | 13 | 16 | 18 |
| Tee or cross (flow turned 90°) | 4 | 5 | 6 | 8 | 10 | 12 | 15 | 17 | 20 | 25 | 30 | 35 | 50 | 60 |
| Butterfly valve | – | – | – | – | 6 | 7 | 10 | – | 12 | 9 | 10 | 12 | 19 | 21 |
| Gate valve | – | – | – | – | 1 | 1 | 1 | 1 | 2 | 2 | 3 | 4 | 5 | 6 |
| Swing check[a] | – | 5 | 7 | 9 | 11 | 14 | 16 | 19 | 22 | 27 | 32 | 45 | 55 | 65 |

[a]Due to the variations in design of swing check valves, the pipe equivalents indicated In the above chart to be considered average



H. S. Wass Jr. and R. P. Fleming P.E., *Sprinkler Hydraulics*,
https://doi.org/10.1007/978-3-030-02595-3

For other than Schedule 40 steel pipe, multiply the figures in the table above by the following factor:

$$Factor = \left(\frac{Actual\ inside\ diameter}{Schedule\ 40\ steel\ pipe\ inside\ diameter}\right)^{4.87}$$

Use with Hazen and Williams's C = 120 only. For other values of C, the figures in table above should be multiplied by the factors indicated below:

| Value of C | 100 | 120 | 130 | 140 | 150 |
|---|---|---|---|---|---|
| Multiplying factor | 0.713 | 1.00 | 1.16 | 1.33 | 1.51 |

NFPA 13 does not offer guidance on alarm check (ACV) or dry pipe (DPV) valves, suggesting that individual values should be obtained for each model. Factory Mutual does suggest some equivalent pipe lengths, for C = 120, which can be used in the absence of better information.

Their suggested values, with a conversion to C = 100 added, are as follows:

| Valve size | 3″ | 3½″ | 4″ | 5″ | 6″ | 8″ |
|---|---|---|---|---|---|---|
| ACV, C = 120, EPL | 13′ | 15′ | 17′ | 21′ | 25′ | 34′ |
| DPV, C = 100, EPL | 9′ | 11′ | 12′ | 15′ | 18′ | 24′ |

*Underground Fire Service Mains*: As discussed on page 59, if the actual internal diameter is known, it should be used in the Hazen-Williams equation to accurately calculate friction loss. When the internal diameter is not known, or the friction loss is fairly small, use of the nominal pipe size in the Hazen-Williams equation is sufficiently accurate. Following are friction-loss constants based upon the nominal pipe size and a Hazen-Williams "C" of 140:

| | | | |
|---|---|---|---|
| 4″ | $5.659 \times 10^{-7}$ | 14″ | $1.268 \times 10^{-9}$ |
| 6″ | $7.856 \times 10^{-8}$ | 16″ | $6.618 \times 10^{-10}$ |
| 8″ | $1.935 \times 10^{-8}$ | 18″ | $3.729 \times 10^{-10}$ |
| 10″ | $6.528 \times 10^{-9}$ | 20″ | $2.232 \times 10^{-10}$ |
| 12″ | $2.687 \times 10^{-9}$ | 24″ | $9.187 \times 10^{-11}$ |

**Table A.2** Internal diameter of sprinkler piping (in.)

| Pipe size | Schedule 40 | Schedule 10 | Pre-1978 thinwall | Type M CU |
|---|---|---|---|---|
| ½″ | 0.622 | 0.674 | | |
| ¾″ | 0.824 | 0.884 | | 0.811 |
| 1″ | 1.049 | 1.097 | | 1.055 |
| 1¼″ | 1.380 | 1.442 | | 1.291 |
| 1½″ | 1.610 | 1.682 | | 1.527 |
| 2″ | 2.067 | 2.157 | | 2.009 |
| 2½″ | 2.469 | 2.635 | | 2.495 |
| 3″ | 3.068 | 3.260 | | 2.981 |
| 3½″ | 3.548 | 3.760 | | 3.459 |
| 4″ | 4.026 | 4.260 | 4.124 | 3.935 |
| 5″ | 5.047 | 5.295 | 5.187 | 4.907 |
| 6″ | 6.065 | 6.357 | 6.249 | 5.881 |
| 8″ | 8.071[a] | 8.249 | | 7.785 |
| 10″ | 10.136[a] | 10.374 | | 9.701 |
| 12″ | 12.090[a] | | | 11.617 |

[a]Schedule 30

**Table A.3** Outside dimensions of sprinkler piping

| Nominal pipe size | Schedule 40 and Schedule 10 | | Type M Copper | |
|---|---|---|---|---|
| | Diameter (in.) | Circumference (in.) | Diameter (in.) | Circumference (in.) |
| ½″ | 0.840 | 2.6 | | |
| ¾″ | 1.050 | 3.3 | 0.875 | 2.7 |
| 1″ | 1.315 | 4.1 | 1.125 | 3.5 |
| 1¼″ | 1.660 | 5.2 | 1.375 | 4.3 |
| 1½″ | 1.900 | 6.0 | 1.625 | 5.1 |
| 2″ | 2.375 | 7.5 | 2.125 | 6.7 |
| 2½″ | 2.875 | 9.0 | 2.625 | 8.2 |
| 3″ | 3.500 | 11.0 | 3.125 | 9.8 |
| 3½″ | 4.000 | 12.6 | 3.625 | 11.4 |
| 4″ | 4.500 | 14.1 | 4.125 | 12.9 |
| 5″ | 5.563 | 17.5 | 5.125 | 16.1 |
| 6″ | 6.625 | 20.8 | 6.125 | 19.2 |
| 8″ | 8.625[a] | 27.1[a] | | |
| 10″ | 10.750[a] | 33.8[a] | | |
| 12″ | 12.750[a] | 40.0[a] | | |

[a]Schedule 30

**Table A.4** Friction loss constants

Friction loss, in psi/linear foot = Friction loss constant × Flow $Q^{1.85}$
FL (psi/ft) = $KQ^{1.85}$

| Pipe size | Schedule 40 C = 120 | Schedule 10 C = 120 | Type M CU C = 150 | Schedule 40 C = 100 | Schedule 10 C = 100 |
|---|---|---|---|---|---|
| ½″ | $6.500 \times 10^{-3}$ | | | $9.107 \times 10^{3}$ | |
| ¾″ | $1.652 \times 10^{-3}$ | | $1.18 \times 10^{-3}$ | $2.315 \times 10^{-3}$ | |
| 1″ | $5.099 \times 10^{-4}$ | $4.101 \times 10^{-4}$ | $3.28 \times 10^{-4}$ | $7.144 \times 10^{-4}$ | $5.746 \times 10^{-4}$ |
| 1¼″ | $1.341 \times 10^{-4}$ | $1.083 \times 10^{-4}$ | $1.23 \times 10^{-4}$ | $1.879 \times 10^{-4}$ | $1.517 \times 10^{-4}$ |
| 1½″ | $6.330 \times 10^{-5}$ | $5.115 \times 10^{-5}$ | $5.42 \times 10^{-5}$ | $8.869 \times 10^{5}$ | $7.167 \times 10^{5}$ |
| 2″ | $1.875 \times 10^{-5}$ | $1.523 \times 10^{-5}$ | $1.42 \times 10^{-5}$ | $2.627 \times 10^{-5}$ | $2.134 \times 10^{-5}$ |
| 2½″ | $7.890 \times 10^{-6}$ | $5.747 \times 10^{-6}$ | $4.96 \times 10^{-6}$ | $1.106 \times 10^{-5}$ | $8.053 \times 10^{-6}$ |
| 3″ | $2.739 \times 10^{-6}$ | $2.038 \times 10^{-6}$ | $2.08 \times 10^{-6}$ | $3.838 \times 10^{-6}$ | $2.856 \times 10^{-6}$ |
| 3½″ | $1.350 \times 10^{-6}$ | $1.017 \times 10^{-6}$ | $1.0\text{I} \times 10^{-6}$ | $1.891 \times 10^{-6}$ | $1.426 \times 10^{-6}$ |
| 4″ | $7.293 \times 10^{-7}$ | $5.539 \times 10^{-7}$ | $5.39 \times 10^{-7}$ | $1.022 \times 10^{-6}$ | $7.761 \times 10^{-7}$ |
| 5″ | $2.426 \times 10^{-7}$ | $1.921 \times 10^{-7}$ | $1.84 \times 10^{-7}$ | $3.399 \times 10^{-7}$ | $2.691 \times 10^{-7}$ |
| 6″ | $9.914 \times 10^{-8}$ | $7.885 \times 10^{-8}$ | $7.62 \times 10^{-8}$ | $1.389 \times 10^{-8}$ | $1.105 \times 10^{-7}$ |
| 8″ | $2.466 \times 10^{-8}$ | $2.217 \times 10^{-8}$ | $1.94 \times 10^{-8}$ | $3.455 \times 10^{-8}$ | $3.106 \times 10^{-8}$ |

To use these tables, go down the left column to the line that represents the actual (nominal) pipe diameter, and go across to the column for the diameter you wish to convert to. Multiply the actual pipe length by the factor thus obtained. For example, suppose you wish to convert 100 feet of 2-in. Schedule 40 pipe to the equivalent length of 2½ in. pipe; that is, the length of 2½ in. pipe, which will produce the same friction loss at any given rate of flow as 100 feet of 2-in. pipe.

**Table A.5** Velocity pressure constants

Velocity pressure = Velocity pressure constant × $Flow^2$
$VP = KQ^2$

| Upstream pipe size | Schedule 40 | Schedule 10 | Type M CU |
|---|---|---|---|
| ¾″ | $2.44 \times 10^{-3}$ | | $7.50 \times 10^{-3}$ |
| 1″ | $9.27 \times 10^{-4}$ | $7.76 \times 10^{-4}$ | $9.07 \times 10^{-4}$ |
| 1¼″ | $3.10 \times 10^{-4}$ | $2.60 \times 10^{-4}$ | $4.04 \times 10^{-4}$ |
| 1½″ | $1.67 \times 10^{-4}$ | $1.40 \times 10^{-4}$ | $2.07 \times 10^{-4}$ |
| 2″ | $6.15 \times 10^{-5}$ | $5.19 \times 10^{-5}$ | $6.89 \times 10^{-5}$ |
| 2½″ | $3.02 \times 10^{-5}$ | $2.33 \times 10^{-5}$ | $2.90 \times 10^{-5}$ |
| 3″ | $1.27 \times 10^{-5}$ | $9.94 \times 10^{-5}$ | $1.42 \times 10^{-5}$ |
| 3½″ | $7.09 \times 10^{-6}$ | $5.62 \times 10^{-6}$ | $7.84 \times 10^{-6}$ |
| 4″ | $4.27 \times 10^{-6}$ | $3.41 \times 10^{-6}$ | $4.68 \times 10^{-6}$ |
| 5″ | $1.73 \times 10^{-6}$ | $1.43 \times 10^{-6}$ | $1.94 \times 10^{-6}$ |
| 6″ | $8.30 \times 10^{-7}$ | $6.88 \times 10^{-7}$ | $9.39 \times 10^{-7}$ |
| 8″ | $2.65 \times 10^{-7}$ | | $3.06 \times 10^{-7}$ |
| 10″ | $1.06 \times 10^{-7}$ | | $1.27 \times 10^{-7}$ |
| 12″ | $5.26 \times 10^{-8}$ | | $6.17 \times 10^{-8}$ |

For conversions not in these tables, use Eq. 16 on page 116 to calculate the factor.

**Table A.6** Equivalent pipe length factors

| | | | | | | | | | | |
|---|---|---|---|---|---|---|---|---|---|---|
| *Schedule 40 pipe* | | | | | | | | | | |
| | 1″ | 1¼″ | 1½″ | 2″ | 2½″ | 3″ | 3½″ | 4″ | 5″ | 6″ |
| 1″ | 3.80 | 8.05 | 27.20 | 64.63 | | | | | | |
| 1¼″ | | 0.263 | 2.12 | 7.15 | 17.00 | 48.95 | | | | |
| 1½″ | 0.124 | 0.472 | | 3.38 | 8,02 | 23.11 | | | | |
| 2″ | 0.0368 | 0.140 | 0.296 | | 2.38 | 6.84 | 13.89 | 25.71 | 77.28 | |
| 2½″ | 0.0155 | 0.0588 | 0.125 | 0.421 | | 2.88 | 5.85 | 10.82 | 32.52 | 79.58 |
| 3″ | | 0.0204 | 0.043 | 0.146 | 0.347 | | 2.0:1 | 3.76 | 11.29 | 27.63 |
| 3½″ | | | | 0.720 | 0.171 | 0.493 | | 1.85 | 5.56 | 13.61 |
| 4″ | | | | 0.0389 | 0.924 | 0.266 | 0.540 | | 3.01 | 7.36 |
| 5″ | | | | 0.0129 | 0.0307 | 0.0880 | 0.180 | 0.333 | | 2.45 |
| 6″ | | | | | 0.0126 | 0.0362 | 0.0735 | 0.136 | 0.409 | |
| *Schedule 10 pipe* | | | | | | | | | | |
| | 1″ | 1¼″ | 1½″ | 2″ | 2½″ | 3″ | 3½″ | 4″ | 5″ | 6″ |
| 1″ | | 3.79 | 8.02 | 26.9 | 71.35 | | | | | |
| 1¼″ | 0.264 | | 2.12 | 7.11 | 18.84 | 53.11 | | | | |
| 1½″ | 0.125 | 0.472 | | 3.36 | 8,90 | 25.10 | | | | |
| 2″ | 0.037 | 0.141 | 0.298 | | 2.65 | 7.47 | 14.97 | 27.50 | 79.32 | |
| 2½″ | O.D140 | 0.0531 | 0.112 | 0.377 | | 2.82 | 5.65 | 10.38 | 29.92 | 72.88 |
| 3″ | | 0.0188 | 0.039 | 0.134 | 0.355 | | 2.00 | 3.68 | 10.61 | 25.85 |
| 3½″ | | | | 0.0668 | 0.171 | 0.499 | | 1.84 | 5.30 | 12.90 |
| 4″ | | | | 0.0363 | 0.0964 | 0.272 | 0.544 | | 2.88 | 7.02 |
| 5″ | | | | 0.0126 | 0.0334 | 0.0942 | 0.189 | 0.347 | | 2.44 |
| 6″ | | | | | 0.0137 | 0.0387 | 0.0775 | 0.142 | 0.411 | |

**Table A.7** Branch line table ordinary hazard, 1–2–3 schedule

| Number of heads on branch line | Distance between heads (Ft.) | | | | | | | | | | |
|---|---|---|---|---|---|---|---|---|---|---|---|
| | 7.5 | 8 | 8.5 | 9 | 9.5 | 10 | 10.5 | 11 | 113 | 12 | 12.5 |
| 2 | 51.4 | 51.6 | 51.8 | 52 | 52.1 | 52.3 | 52.5 | 52.6 | 52.8 | 53 | 53.2 |
| | 24.7 | 25 | 25.3 | 25.7 | 26 | 26.3 | 26.6 | 26.9 | 27.3 | 27.6 | 27.9 |
| 3 | 81.2 | 81.7 | 82.3 | 82.8 | 83.4 | 83.9 | 84.4 | 85 | 85.5 | 86 | 86.5 |
| | 30.3 | 31 | 31.8 | 32.5 | 33.3 | 34 | 34.8 | 35.5 | 36.3 | 37 | 37.8 |
| 4 | 112.7 | 113.7 | 114.7 | 115.7 | 116.7 | 117.7 | 118.7 | 119.7 | 120.7 | 121.7 | 122.7 |
| | 33.7 | 34.7 | 35.8 | 36.8 | 37.8 | 38.9 | 39.9 | 41 | 42.1 | 43.1 | 44.2 |
| 5 | 144.9 | 146.5 | 148.1 | 149.7 | 151.3 | 152.8 | 154.4 | 155.9 | 157.5 | 159 | 160.6 |
| | 36.7 | 38 | 39.2 | 40.5 | 41.8 | 43.2 | 44.5 | 45.9 | 47.2 | 48.6 | 50 |
| 6 | 180.1 | 182.4 | 184.7 | 187 | 189.3 | 191.6 | 193.8 | 196.1 | 198.4 | 200.6 | 202.8 |
| | 41.4 | 43.1 | 44.8 | 46.5 | 48.3 | 50.1 | 51.9 | 53.8 | 55.7 | 57.6 | 59.6 |
| 7 | 215.8 | 218.9 | 222 | 225 | 228.1 | 231.1 | 234.2 | 237.2 | 240.2 | 243.3 | 246.3 |
| | 43.5 | 45.4 | 47.3 | 49.2 | 51.2 | 53.2 | 55.3 | 57.4 | 59.5 | 61.7 | 63.9 |
| 8 | 252.3 | 256.3 | 260.2 | 264.1 | 268 | 271.9 | 275.8 | 279.7 | 283.6 | 287.5 | 291.4 |
| | 46.5 | 48.6 | 50.8 | 53 | 55.3 | 57.7 | 60.1 | 62.5 | 65 | 67.5 | 70.1 |
| Each pair of numbers:<br>Flow (GPM)<br>Pressure (PSI) | | | | | | End head row = 25.0 GP<br>Schedule 40 Pipe<br>C = 120<br>k = 5.6 | | | | | |

**Table A.8** Branch line table ordinary hazard 2–3–5 schedule

| Number of heads on branch line | Distance between heads (Ft.) | | | | | | | | | | |
|---|---|---|---|---|---|---|---|---|---|---|---|
| | 77.5 | 8 | 8.5 | 9 | 9.5 | 10 | 10.5 | 11 | 11.5 | 12 | 12.5 |
| 2 | 449.5 | 49.6 | 49.6 | 49.7 | 49.7 | 49.8 | 49.8 | 49.9 | 50 | 50 | 50.1 |
| | 221.4 | 21.5 | 21.6 | 21.7 | 21.8 | 21.9 | 22 | 22.1 | 22.2 | 22.3 | 22.4 |
| 3 | 777.3 | 77.6 | 77.9 | 78.2 | 78.4 | 78.8 | 79 | 79.4 | 79.7 | 80 | 80.3 |
| | 226.6 | 27.1 | 27.5 | 28 | 28.5 | 28.9 | 29.4 | 29.9 | 30.3 | 30.8 | 31.3 |
| 4 | 1106.8 | 107.5 | 108.1 | 108.8 | 109.3 | 110.1 | 110.6 | 111.4 | 112 | 112.7 | 113.3 |
| | 2<br>229.8 | 30.5 | 31.1 | 31.9 | 32.5 | 33.3 | 34 | 34.7 | 35.4 | 36.1 | 36.9 |
| 5 | 137.1 | 138.3 | 139.3 | 140.4 | 141.3 | 142.6 | 143.5 | 144.8 | 145.8 | 146.9 | 147.9 |
| | 32.4 | 33.4 | 34.3 | 35.2 | 36.1 | 37.1 | 38 | 39 | 39.9 | 40.9 | 41.9 |
| 6 | 7170.2 | 172 | 173.5 | 175.2 | 176.7 | 178.6 | 180 | 181.9 | 183.4 | 185.1 | 186.7 |
| | 336.7 | 38 | 39.3 | 40.5 | 41.8 | 43.2 | 44.5 | 45.9 | 47.3 | 48.7 | 50.1 |
| 7 | 2203.8 | 206.2 | 208.4 | 210.7 | 212.8 | 215.4 | 217.3 | 219.9 | 222 | 224.3 | 226.5 |
| | 338.6 | 40 | 41.5 | 42.9 | 44.4 | 45.9 | 47.4 | 49 | 50.6 | 52.2 | 53.8 |
| 8 | 238.2 | 241.3 | 244.2 | 247.2 | 250 | 253.3 | 255.9 | 259.2 | 262 | 265 | 267.9 |
| | 41.2 | 42.9 | 44.6 | 46.3 | 48 | 49.8 | 51.6 | 53.5 | 55.3 | 57.2 | 59.1 |
| Each pair of numbers:<br>Flow (GPM)<br>Pressure (PSI) | | | | | End head flow = 250 GPM<br>Schedule 40 Pipe<br>C=120<br>k=5.6 | | | | | | |

**Table A.9** Branch line table extra hazard schedule

| Number of heads on branch line | Distance between heads (Ft.) | | | | | | | | | | |
|---|---|---|---|---|---|---|---|---|---|---|---|
| | 77 | 7.5 | 8 | 8.5 | 9 | 9.5 | 10 | 10.5 | 11 | 11.5 | 12 |
| 2 | 50.4 | 50.4 | 50.5 | 50.5 | 50.6 | 50.7 | 50.7 | 50.8 | 50.8 | 50.9 | 51 |
| | 21.3 | 21.4 | 21.5 | 21.6 | 21.7 | 21.8 | 21.9 | 22 | 22.1 | 22.2 | 22.3 |
| 3 | 76.4 | 76.6 | 76.8 | 76.9 | 77.1 | 77.3 | 77.5 | 77.6 | 77.8 | 78 | 78.2 |
| | 22.6 | 22.8 | 23 | 23.2 | 23.4 | 23.6 | 23.8 | 24 | 24.2 | 24.4 | 24.6 |
| 4 | 102.8 | 103.2 | 103.5 | 103.8 | 104.2 | 104.5 | 104.9 | 105.2 | 105.6 | 105.9 | 106.2 |
| | 24 | 24.3 | 24.6 | 24.9 | 25.2 | 25.5 | 25.8 | 26.1 | 26.4 | 26.7 | 27 |
| 5 | 1130 | 130.6 | 131.2 | 131.8 | 132.4 | 133 | 133.6 | 134.2 | 134.8 | 135.4 | 136 |
| | 26.3 | 26.8 | 27.3 | 27.8 | 28.3 | 28.8 | 29.3 | 29.8 | 30.3 | 30.8 | 31.3 |
| 6 | 159.8 | 160.8 | 161.8 | 162.8 | 163.8 | 164.7 | 165.7 | 166.7 | 167.7 | 168.7 | 169.7 |
| | 229.9 | 30.7 | 31.5 | 32.3 | 33.1 | 33.9 | 34.7 | 35.5 | 36.3 | 37.2 | 38.0 |
| Each pair of numbers: Flow (GPM)<br>Pressure (PSI) | | | | | End head flow: 25.0 GPM<br>Schedule 40 Pipe<br>C = 120<br>k = 5.6 | | | | | | |

**Table A.10** Branch line table extra hazard schedule with large orifice heads

| Number of heads on branch line | Distance between heads (Ft.) | | | | | | | | | | |
|---|---|---|---|---|---|---|---|---|---|---|---|
| | 77 | 7.5 | 8 | 8.5 | 9 | 9.5 | 10 | 10.5 | 11 | 11.5 | 12 |
| 2 | 50.8 | 50.9 | 51 | 51.1 | 51.2 | 51.3 | 51.5 | 51.6 | 51.7 | 51.8 | 51.9 |
| | 10.9 | 11 | 11.1 | 11.2 | 11.3 | 11.4 | 11.5 | 11.6 | 11.7 | 11.8 | 11.9 |
| 3 | 77.9 | 78.3 | 78.6 | 78.9 | 79.3 | 79.6 | 80 | 80.3 | 80.7 | 81 | 81.3 |
| | 12.2 | 12.4 | 12.6 | 12.9 | 13.1 | 13.3 | 13.5 | 13.7 | 13.9 | 14.1 | 14.3 |
| 4 | 105.7 | 106.4 | 107.1 | 107.8 | 108.4 | 109.1 | 109.8 | 110.5 | 111.1 | 111.8 | 112.5 |
| | 13.6 | 14 | 14.3 | 14.6 | 14.9 | 15.2 | 15.6 | 15.9 | 16.2 | 16.6 | 16.9 |
| 5 | 135 | 136.2 | 137.4 | 138.6 | 139.8 | 141 | 142.1 | 143.3 | 144.5 | 145.7 | 146.9 |
| | 16.1 | 16.6 | 17.2 | 17.7 | 18.2 | 18.8 | 19.3 | 19.9 | 20.5 | 21 | 21.6 |
| 6 | 169.5 | 171.5 | 173.5 | 175.5 | 177.4 | 179.4 | 181.3 | 183.3 | 185.2 | 187.1 | 189.1 |
| | 20 | 20.8 | 21.7 | 22.6 | 23.5 | 24.5 | 25.4 | 26.4 | 27.4 | 28.4 | 29.4 |
| Each pair of numbers:<br>Flow (GPM)<br>Pressure (PSI) | | | | | | End head flows 25.0 GPM<br>Schedule 40 Pipe<br>C = 120<br>k = 8.1 | | | | | |

# Appendix B
# Summary of Useful Equations

Definition of Recurring Symbols
Q = flow, in gallons per minute (gpm)
p = pressure, in pounds per square inch (psi)
d = internal diameter of pipe or orifice (in.)
L = length (feet)
*See Appendix C for SI version. All other equations may be used as shown in this Appendix with appropriate SI values.

*1. Flow Through an Orifice (page 42)

$$Q = 29.84cd^2\sqrt{p} \qquad \text{c = discharge coefficient}$$

2. Flow from a Sprinkler (page 42)

$$Q = k\sqrt{p}$$

k = discharge coefficient (see Appendix D for SI conversion factor)

*3. Hazen-Williams Friction Loss Equation (page 53)

$$p = \frac{4.52}{C^{1.85}d^{4.87}} \times Q^{1.85}$$

P friction loss per foot of pipe, psi
C Hazen-Williams coefficient



H. S. Wass Jr. and R. P. Fleming P.E., *Sprinkler Hydraulics*,
https://doi.org/10.1007/978-3-030-02595-3

*4. Hazen-Williams "C," When Friction Loss Is Known (page 51)

$$C = \left(\frac{6.05 \times 10^5}{pd^{4.87}}\right)^{0.54} \times Q$$

p friction loss, in psi, per foot of pipe

5. Normal Pressure (page 74)

$$P_N = P_T - P_v$$

$P_N$ normal pressure
$P_T$ total pressure
$P_v$ velocity pressure

*6. Velocity Pressure (page 75)

$$P_v = 0.001123\frac{Q^2}{D^4}$$

*7. Flow Velocity in a Pipe (page 86)

$$v = \frac{0.4085Q}{d^2}$$

v flow velocity, in feet per second

8. Approximating Flow at a Given Pressure When Flow Is Known for a Different Pressure (page 90)

$$Q_2 = Q_1\sqrt{\frac{P_2}{P_1}}$$

$Q_2$ flow to be approximated
$Q_1$ flow associated with known pressure
$P_2$ pressure for which associated approximate flow is to be determined
$P_1$ known pressure associated with known flow

9. Flow at Junction Point Between a Sprinkler Demand Curve and the Water-Supply Curve (page 105)

$$Q_J = \left[\frac{P_S - P_E}{\frac{P_S - P_R}{Q_F^{1.85}} + \frac{P_D - P_E}{Q_D^{1.85}}}\right]^{0.54}$$

$Q_J$ flow at junction point
$Q_F$ flow in flow test
$Q_D$ design flow
$P_R$ residual pressure, flow test
$P_S$ static pressure, flow test
$P_D$ design pressure, at effective point of flow test
$P_E$ height of design area, psi (0.433 × height in feet)

10. Pressure at Junction Point Between a Sprinkler Demand Curve and the Water-Supply Curve, After Using Eq. 9 (page 105)

$$P_J = P_S - \left(\frac{P_S - P_R}{Q_F^{1.85}}\right) Q_J^{1.85}$$

$P_J$ pressure at junction point

11. Flow at any Specified Pressure on a Water-Supply Curve (page 106)

$$Q = Q_F - \left(\frac{P_S - P}{P_S - P_R}\right)^{0.54}$$

P specified pressure
$P_S$ static pressure, flow test
$P_R$ residual pressure, flow test
$Q_F$ flow in flow test

12. Pressure at any Specified Flow on a Water-Supply Curve (page 106)

$$P = P_S - \left(\frac{Q}{Q_F}\right)^{1.85} \times (P_S - P_R)$$

Q specified flow

See Eq. 11 for other symbols

13. Flow at Junction Point Between Water-Supply Curve, Less Specified Hose-Stream Allowance, and Sprinkler, Demand Curve (Solution involves trial and error.) (page 106)

$$\frac{P_P - P_E}{Q_D^{1.85}} \times Q^{1.85} + \frac{P_S - P_R}{Q_F^{1.85}} \times Q^{1.85} \times (Q + H)^{1.85} = P_S - P_E$$

Q flow at junction point
H hose-stream allowance, in gpm See Eq. 9 for other symbols

14. After Determining "Q" in Eq. 13, Associated Pressure (page 106)

$$P = P_E + \frac{P_D - P_E}{Q_D^{1.85}} \times Q^{1.85}$$

15. Flow Through One Leg of a Loop, All Pipe Size the Same (page 115)

$$Q_1 = \frac{Q}{\left(\frac{L_2}{L_1}\right)^{0.54} + 1}$$

| | |
|---|---|
| $L_1$ and $L_2$ | are the two legs of the loop |
| Q | total flow through the loop |
| $Q_1$ | flow through leg $L_1$ |

16. Equivalent (in Terms of Friction Loss) Pipe Length Factor (page 116)

$$FACTOR = \left(\frac{D_2}{D_1}\right)^{4.87}$$

$d_1$ actual internal diameter
$d_2$ internal diameter of pipe being converted to

17. Length of Pipe, of Same Internal Diameter, Which Has Same Friction Loss as Two Legs in Parallel (page 117)

$$Equivalent\,Length = L_E = \left[\frac{1}{\left(\frac{L_1}{L_2}\right)^{0.54} + 1}\right]^{1.85} \times L_1$$

18. Hardy Cross Equation (page 135)

$$\Delta = -\frac{\Sigma k Q_A^{1.85}}{1.85 \Sigma\, k Q_A^{0.85}}$$

k is a friction loss constant

$Q_A$ is an assumed flow

# Appendix C
# SI Version of Equations in Appendix B

*Note* Where SI version of equation is not shown here, the equation in Appendix B can be used with SI values.

Definition of Recurring Symbols
Q = flow, in liters per minute (L/min)
p = pressure, in bars
d = internal diameter of pipe or orifice, in millimeters (mm)

1. Flow through an orifice (page 36)

$$Q = 0.6667cd^2\sqrt{p}$$

c discharge coefficient

2. Hazen-Williams Friction Loss Equation (page 48)

$$P = \frac{6.05 \times 10^5}{C^{1.85}d^{4.87}} \times Q^{1.85}$$

p friction loss per meter of pipe, bars
C Hazen-Williams coefficient

3. Hazen-Williams "C", when friction loss is known (page 51)

$$C = \left(\frac{6.05x10^5}{pd^{4.87}}\right)^{0.54} \times Q$$

p friction loss per meter of a pipe, bars



H. S. Wass Jr. and R. P. Fleming P.E., *Sprinkler Hydraulics*,
https://doi.org/10.1007/978-3-030-02595-3

4. Velocity Pressure (page 77)

$$P = 2.25158\frac{Q^2}{d^4}$$

5. Flow Velocity in a Pipe (page 77)

$$v = 21.22\frac{Q}{d^2}$$

v flow velocity, in meters per second

# Appendix D
# Conversion Factors Between U.S. and SI Units of Measurement

## Length

| | | | |
|---|---|---|---|
| inch | millimeter | mm | 25.400 mm |
| inch | centimeter | cm | 2.540 cm |
| foot | meter | m | 0.3048 m |
| millimeter | inch | i or in. | 0.03937 i |
| centimeter | inch | i or in. | 0.3937 i |
| meter | foot | ft. | 3.281 ft. |

## Area

| | | | |
|---|---|---|---|
| square inch | square millimeter | $mm^2$ | 645.16 $mm^2$ |
| square foot | square meter | $m^2$ | 0.0929 $m^2$ |
| sq. millimeter | square inch | $in.^2$ | 0.00155 $in.^2$ |
| square meter | square foot | $ft^2$ | 10.764 sq. ft. |

## Volume

| | | | |
|---|---|---|---|
| gallon | liter | L | 3.785 L |
| liter | gallon | g | 0.264 g |

## Flow Rate

| | | | |
|---|---|---|---|
| gallons/min | liters/minute | L/min | 3.785 L/min |
| gallons/min | cubic meters/min | $m^3$/min | 0.00379 $m^3$/min |
| liters/min | gallons/minute | gpm | 0.264 gpm |
| cubic meters/min | gallons/minute | gpm | 264.2 gpm |



H. S. Wass Jr. and R. P. Fleming P.E., *Sprinkler Hydraulics*,
https://doi.org/10.1007/978-3-030-02595-3

## Water Density

| | | |
|---|---|---|
| gallons/min/ft$^2$ | liters/min/meter$^2$ | 40.746 L/min/m$^2$ |
| liters/min/meter | gallons/min/sq. ft. | 0.0245 gpm/ft$^2$ |

Liters per minute per square meter is the metric equivalent of gallons per minute per square foot and we express it that way because the relationship is obvious. Because of the simple relationship between liters and meters, the rest of the world has a simpler way to express "liters per square meter."

Density is a volume divided by an area over a unit of time. 1 L = 1,000 cm$^3$ or 1,000,000 mm$^3$. 1 m = 1,000 mm. Thus 1 m$^2$ = 1,000,000 mm$^2$. Dividing volume (1,000,000 mm$^3$) by area (1,000,000 mm$^2$) we get 1 mm. Therefore, 1 L/min/m$^2$ = 1 mm/min.

For the first time, the 1999 Edition of NFPA 13 ma y refer to metric density equivalents as "millimeters (mm) per minute". I qualify this because it did not appear in the "Proposal" and "Comment" texts and the 1999 Edition has not been published as this goes to press.

Unfortunately, the relationship in U.S. units is not as arithmetically simple. Whereas liters per square meter equals millimeters, a conversion factor of 1.604 must be applied to gallons per square foot to convert to inches. Thus a density of 0.30 gpm/ft$^2$/min is equal to about 0.48 in./min. Because of this conversion, inches per minute is not likely to gain favor.

At first glance, describing density in terms of millimeters or inches per minute may seem strange. In the United States we are, however, used to rainfall being described in inches. We may read about a severe thunderstorm where 2 in. of rain was measured in 20 min. If the rate of rainfall remained constant during the 20 min time span (unlikely), Mother Nature would have been delivering a density of 0.10 in./min, or about 0.062 gpm/ft$^2$/min.

## Pressure

| | | |
|---|---|---|
| pounds per sq in. | pascal Pa | 6.895 kPa |
| pounds per sq in. | bar | 0.06895 bar |
| kilopascal | pounds/sq in. psi | 0.145 psi |
| bar | pounds/sq in. psi | 14.503 psi |

## Head

1 ft. water = 0.433 psi
1 m water = 0.0980 bar
1 m water = 9.802 Kpa

## Temperature

degrees C = 5/9 × (degrees F − 32)
degrees F = (9/5 × degrees C) + 32

**Sprinkler K-factor**

If pressure is measured in bars: multiply the K-factor for English units by 14.414. The resultant flow will be in liters per minute.

If pressure is measured in kilopascals: multiply the K-factor for English units by 1.441. The resultant flow will be in liters per minute.

**Nominal Pipe and Tube Sizes**

| inches | millimeters |
|---|---|
| ¾ | 19 |
| 1 | 25 |
| 1-¼ | 31 |
| 1-½ | 37 |
| 2 | 50 |
| 2-½ | 62 |
| 3 | 75 |
| 3-½ | 87 |
| 4 | 100 |
| 5 | 125 |
| 6 | 150 |
| 8 | 200 |
| 10 | 250 |
| 12 | 300 |

*Note* Occasionally you will find the nominal diameter in millimeters expressed in slightly different numbers than those shown above.

**Response Time Index (RTI)**

RTI units in the English system are seconds$^{½}$ feet$^{½}$

RTI units in the SI system are seconds$^{½}$ meters$^{½}$

RTI in the SI system = 0.552 RTI in the English system. RTI in the English system = 1.811 RTI in the SI system.

# Appendix E
# Friction Loss Table

On the following pages is a table that can be used to determine friction loss, in psi per foot of pipe, for Schedule 40 sprinkler piping where C = 120.

The numbers beneath the pipe size represent the flows, in gpm, which produce the indicated friction loss. To use the table, simply select the column for the pipe size you are concerned with and go down the column until you reach the approximate flow you are looking for and read the friction loss in the "PSI per ft." column to the left. Although for most calculations the additional accuracy is meaningless, you may make a linear interpolation if you wish.

This friction-loss table can be used for other "C's" and other than Schedule 40 pipe by multiplying the friction loss derived from the table by the following modification factors:

| Pipe size | Standard underground C = 140 | Schedule 40 C = 100 | Schedule 10 C = 120 | Type M Schedule 10 C = 100 | Copper C = 150 | Pre-1978 Schedule 10 C = 120 | Pre-1978 Schedule 10 C = 100 |
|---|---|---|---|---|---|---|---|
| ¾ | | 1.40 | | | | 0.715 | |
| 1 | | 1.40 | 0.804 | 1.127 | 0.644 | | |
| 1¼ | | 1.40 | 0.807 | 1.131 | 0.916 | | |
| 1½ | | 1.40 | 0.808 | 1.132 | 0.856 | | |
| 2 | | 1.40 | 0.883 | 1.138 | 0.760 | | |
| 2½ | | 1.40 | 0.728 | 1.021 | 0.629 | | |
| 3 | | 1.40 | 0.744 | 1.043 | 0.761 | | |
| 3½ | | 1.40 | 0.754 | 1.056 | 0.749 | | |
| 4 | 0.752 | 1.40 | 0.759 | 1.064 | 0.740 | 0.889 | 1.246 |
| 5 | 0.752 | 1.40 | 0.792 | 1.109 | 0.759 | 0.875 | 1.226 |
| 6 | 0.752 | 1.40 | 0.795 | 1.114 | 0.769 | 0.864 | 1.211 |

(continued)



H. S. Wass Jr. and R. P. Fleming P.E., *Sprinkler Hydraulics*,
https://doi.org/10.1007/978-3-030-02595-3

(continued)

| Pipe size | Standard underground C = 140 | Schedule 40 C = 100 | Schedule 10 C = 120 | Type M Schedule 10 C = 100 | Copper C = 150 | Pre-1978 Schedule 10 C = 120 | Pre-1978 Schedule 10 C = 100 |
|---|---|---|---|---|---|---|---|
| 8 | 0.752 | 1.40 | | | | | |
| 10 | 0.752 | 1.40 | | | | | |
| 12 | 0.752 | 1.40 | | | | | |

## Modification factors for pipe in Appendix G

| Nom. pipe size | Allied “XL” | | “Poz-Lock” | |
|---|---|---|---|---|
| | C = 120 | C = 100 | C = 120 | C = 100 |
| 1 | 0.711 | 0.997 | 1.238 | 1.735 |
| 1¼ | 0.723 | 1.013 | 1.772 | 2.483 |
| 1½ | 0.746 | 1.045 | 1.166 | 1.634 |
| 2 | 0.777 | 1.088 | | |
| 2½ | 0.767 | 1.075 | | |
| 3 | 0.783 | 1.097 | | |

| Nom. pipe size | Central Sprinkler “TL” Threadable Lightwall | | American Tube DynaThread 40 | |
|---|---|---|---|---|
| | C = 120 | C = 100 | C = 120 | C = 100 |
| 1 | 0.794 | 1.064 | 0.841 | 1.178 |
| 1¼ | 0.765 | 1.072 | 0.882 | 1.736 |
| 1½ | 0.783 | 1.097 | 0.887 | 1.243 |
| 2 | 0.867 | 1.131 | 0.894 | 1.253 |

| Nom. pipe size | American Tube Black Lightwall Threadable (BLT) | | American Tube Dyna Flow 10 | |
|---|---|---|---|---|
| | C = 120 | C = 100 | C = 120 | C = 100 |
| 1 | 0.780 | 1.092 | 0.526 | 0.737 |
| 1¼ | 0.781 | 1.094 | 0.582 | 0.816 |
| 1½ | 0.797 | 1.116 | 0.663 | 0.928 |
| 2 | 0.818 | 1.46 | 0.696 | 0.975 |
| 2½ | | | 0.612 | 0.857 |
| 3 | | | 0.661 | 0.926 |
| 4 | | | 0.703 | 0.985 |

| Nom. pipe size | American Tube Dyna Light-S | | Polybutylene CTS | Polybutylene IPS |
|---|---|---|---|---|
| | C = 120 | C = 100 | C = 150 | C = 150 |
| 1 | 0.552 | 0.774 | 1.315 | 0.656 |
| 1¼ | 0.605 | 0.848 | 1.894 | 0.786 |
| 1½ | 0.630 | 0.883 | 1.780 | 0.854 |
| 2 | 0.669 | 0.937 | 1.620 | 0.955 |

| Nom. pipe size | Post-Chlorinated Polyvinyl Chloride (CPVC) | Type L Copper |
|---|---|---|
| | C = 150 | C = 150 |
| 1 | 0.505 | 0.741 |
| 1¼ | 0.617 | 1.010 |
| 1½ | 0.678 | 0.919 |
| 2 | 0.771 | 0.806 |
| 2½ | 0.725 | 0.667 |
| 3 | 0.800 | 0.808 |
| 3½ | | 0.786 |
| 4 | | 0.768 |
| 5 | | 0.784 |
| 6 | | 0.792 |
| 8 | | 0.819 |

## Central Sprinkler "Schedule 7"

| Nom. pipe size | C = 120 | C = 100 |
|---|---|---|
| 1¼ | 0.663 | 0.928 |
| 2 | 0.696 | 0.974 |
| 2½ | 0.653 | 0.914 |
| 3 | 0.681 | 0.953 |
| 4 | 0.726 | 1.016 |

## Friction-Loss Table

| PSI PER FT. | ¾" | 1" | 1¼" | 1½" | 2" | 2½" | PSI PER FT. | ¾" | 1" | 1¼" | 1½" | 2" | 2½" |
|---|---|---|---|---|---|---|---|---|---|---|---|---|---|
| 0.005 | | 3.43 | 7.07 | 10.61 | | | 0.245 | | 28.16 | 57.96 | 86.96 | 167.87 | 268.01 |
| 0.01 | | 5 | 10.28 | 15.43 | 29.79 | 47.56 | 0.25 | | 28.46 | 58.59 | 87.92 | 189.72 | 270.96 |
| 0.015 | 3.29 | 6.22 | 12.8 | 19.21 | 37.09 | 59.22 | 0.255 | | 28.77 | 59.22 | 88.86 | 171.84 | 273.87 |
| 0.02 | | 7.27 | 14.96 | 22.45 | 43.33 | 69.18 | 0.26 | | 29.07 | 59.85 | 89.8 | 173.35 | 276.76 |
| 0.025 | | 8.2 | 16.88 | 25.32 | 48.89 | 78.05 | 0.265 | | 29.38 | 60.47 | 90.73 | 175.15 | 279.63 |
| 0.03 | | 9.05 | 18.62 | 27.95 | 53.95 | 86.13 | 0.27 | | 29.67 | 61.08 | 91.65 | 176.93 | 282.46 |
| 0.035 | | 9.83 | 20.24 | 30.37 | 58.64 | 93.61 | 0.275 | | 29.97 | 61.69 | 92.56 | 178.69 | 285.28 |
| 0.04 | | 10.57 | 21.76 | 32.65 | 63.03 | 100.62 | 0.28 | | 30.26 | 62.29 | 93.47 | 180.44 | 288.07 |
| 0.045 | | 11.27 | 23.19 | 34.79 | 67.17 | 107.24 | 0.285 | | 30.55 | 62.89 | 94.37 | 182.17 | 290.84 |
| 0.05 | | 11.93 | 24.55 | 36.63 | 71.11 | 113.52 | 0.29 | | 30.84 | 63.49 | 95.26 | 163.89 | 293.59 |
| 0.055 | | 12.56 | 25.85 | 38.78 | 74.86 | 119.52 | 0.295 | | 31.13 | 64.08 | 96.14 | 185.6 | 296.31 |
| 0.06 | | 13.16 | 27.09 | 40.65 | 78.47 | 125.28 | 0.3 | 16.64 | 31.41 | 64.66 | 97.02 | 187.29 | 299.02 |
| 0.065 | | 13.74 | 28.29 | 42.45 | 81.94 | 130.82 | 0.305 | 16.79 | 31.69 | 65.24 | 97.89 | 188.98 | 301.7 |
| 0.07 | | 14.3 | 29.44 | 44.18 | 85.29 | 136.16 | 0.31 | 16.94 | 31.97 | 65.82 | 98.76 | 190.64 | 304.37 |
| 0.075 | | 14.85 | 30.56 | 45.86 | 88.53 | 141.34 | 0.315 | 17.08 | 32.25 | 66.39 | 99.61 | 192.3 | 307.01 |
| 0.08 | | 15.38 | 31.65 | 47.49 | 91.67 | 146.36 | 0.32 | 17.23 | 32.53 | 66.96 | 100.47 | 193.94 | 309.63 |
| 0.085 | | 15.89 | 32.7 | 49.07 | 94.73 | 151.23 | 0.325 | 17.37 | 32.8 | 67.52 | 101.31 | 195.58 | 312.24 |
| 0.09 | | 16.39 | 33.73 | 50.61 | 97.7 | 155.98 | 0.33 | 17.52 | 33.07 | 68.08 | 102.15 | 197.2 | 314.63 |
| 0.095 | | 18.87 | 34.73 | 52.15 | 100.6 | 160.6 | 0.335 | 17.66 | 33.34 | 68.63 | 102.98 | 198.81 | 317.4 |
| 0.1 | 9.19 | 17.35 | 35.71 | 53.58 | 103.42 | 165.12 | 0.340 | 17.8 | 33.61 | 69.19 | 103.81 | 200.4 | 319.95 |
| 0.105 | | 17.81 | 36.66 | 55.01 | 106.19 | 169.53 | 0.345 | 17.97: | 33.88 | 69.73 | 104.64 | 201.99 | 322.48 |
| 0.11 | | 18.26 | 37.59 | 56.41 | 108.89 | 173.85 | 0.35 | 18.08 | 34.14 | 70.28 | 105.45 | 203.57 | 325 |
| 0.115 | | 18.71 | 38.51 | 57.78 | 111.$4 | 178.08 | 0.355 | 18.22 | 34.4 | 70.82 | 106.26 | 205.14 | 327.5 |
| 0.12 | | 19.14 | 39.4 | 59.12 | 114.14 | 182.22 | 0.36 | 18.36 | 34.67 | 71.36 | 107.07 | 206.69 | 329.99 |
| 0.125 | | 19.57 | 40.28 | 60.44 | 116.68 | 186.29 | 0.365 | 18.5 | 34.93 | 71.89 | 107.87 | 208.24 | 332.46 |
| 0.13 | | 19.99 | 41.15 | 61.74 | 119.18 | 190.28 | 0.37 | 18.64 | 35.18 | 72.42 | 108.67 | 209.78 | 334.91 |
| 0.135 | | 20.4 | 41.99 | 63.01 | 121.64 | 194.2 | 0.375 | 18.77 | 35.44 | 72.95 | 109.46 | 211.3 | 337.46 |
| 0.14 | | 20.81 | 42.83 | 64.26 | 124.05 | 198.05 | 0.38 | 18.91 | 35.69 | 73.47 | 110.25 | 212.82 | 339.77 |
| 0.145 | | 21.2 | 43.65 | 65.49 | 126.43 | 201.85 | 0.385 | 19.04 | 35.95 | 73.99 | 111.03 | 214.33 | 342.18 |
| 0.150 | | 21.6 | 44.45 | 66.7 | 128.77 | 205.58 | 0.39 | 19.17 | 36.2 | 74.51 | 111.8 | 215.63 | 344.58 |
| 0.155 | | 21.98 | 45.25 | 67.9 | 131.07 | 209.26 | 0.395 | 19.31 | 36.45 | 75.03 | 112.58 | 217.32 | 346.96 |
| 0.160 | | 22.36 | 46.03 | 69.07 | 133.34 | 212.88 | 0.4 | 19.44 | 36.7 | 75.$4 | 113.35 | 218.81 | 349.33 |
| 0.165 | | 22.74 | 46.8 | 70.23 | 135.58 | 216.45 | 0.405 | 19.57 | 36.94 | 76.05 | 114.11 | 220.28 | 351.68 |
| 0.170 | | 23.11 | 47.57 | 71.37 | 137.78 | 219.97 | 0.41 | 19.71 | 37.19 | 76.55 | 114.87 | 221.75 | 3&4.02 |
| 0.175 | | 23.47 | 48.32 | 72.5 | 139.96 | 223.44 | 0.415 | 19.83 | 37.44 | 77.06 | 115.62 | 223.2 | 356.25 |
| 0.180 | | 23.63 | 49.06 | 73.61 | 142.1 | 226.87 | 0.42 | 19.96 | 37.68 | 77.56 | 116.37 | 224.65 | 358.66 |
| 0.185 | | 24.19 | 49.79 | 74.71 | 144.22 | 230.26 | 0.425 | 20.08 | 37.92 | 78.06 | 117.12 | 226.1 | 360.96 |
| 0.190 | | 24.54 | 50.51 | 75.8 | 146.32 | 233.6 | 0.43 | 20.21 | 38.16 | 78.55 | 117.86 | 227.53 | 363.25 |
| 0.195 | | 24.89 | 51.23 | 76.87 | 148.39 | 236.9 | 0.435 | 20.34 | 38.4 | 79.04 | 118.6 | 228.96 | 365.53 |
| 0.2 | 13.36 | 25.23 | 51.93 | 77.93 | 150.43 | 240.17 | 0.44 | 20.46 | 38.64 | 79.53 | 119.34 | 230.37 | 367.8 |
| 0.2 | | 25.57 | 52.63 | 78.97 | 152.45 | 243.39 | 0.445 | 20.59 | 38.87 | 80.02 | 120.07 | 231.79 | 370.05 |
| 0.21 | | 25.9 | 53.32 | 80.01 | 154.45 | 246.59 | 0.45 | 20.71 | 39.11 | 80.5 | 120.8 | 233.19 | 372.29 |
| 0.215 | | 26.24 | 54 | 81.03 | 156.43 | 249.74 | 0.455 | 20.84 | 39.34 | 80.99 | 121.52 | 234.59 | 374.52 |
| 0.22 | | 26.56 | $4.68 | 82.05 | 158.39 | 252.86 | 0.46 | 20.96 | 39.58 | 81.47 | 122.24 | 235.98 | 376.74 |
| 0.225 | | 26.89 | 55.35 | 83.05 | 160.32 | 255.95 | 0.465 | 21.09 | 39.81 | 81.94 | 122.96 | 237.36 | 378.95 |
| 0.23 | | 27.21 | 56.01 | 84.04 | 162.24 | 259.01 | 0.47 | 21.21 | 40.04 | 82.42 | 123.67 | 238.74 | 381.15 |
| 0.235 | | 27.53 | 56.66 | 85.02 | 164.13 | 262.04 | 0.475 | 21.33 | 40.27 | 82.89 | 124.38 | 240.11 | 363.33 |
| 0.24 | | 27.84 | 57.31 | 86 | 166.01 | 265.04 | 0.48 | 21.45 | 40.5 | 83.36 | 125.08 | 241.47 | 385.51 |

# Friction-Loss Table

| PSI PER FT. | ¾″ | 1″ | 1¼″ | 1½″ | 2″ | 2½″ | PSI PER FT. | ¾″ | 1″ | 1¼″ | 1½″ | 2″ | 2½″ |
|---|---|---|---|---|---|---|---|---|---|---|---|---|---|
| 0.485 | 21.57 | 40.73 | 83.83 | 125.79 | 242.82 | 387.67 | 0.725 | 26.81 | 50.61 | 104.18 | 156.32 | 301.77 | 481.78 |
| 0.490 | 21.69 | 40.95 | 84.30 | 126.49 | 244.17 | 389.83 | 0.730 | 26.91 | 50.80 | 104.57 | 156.90 | 302.89 | 483.57 |
| 0.495 | 21.81 | 41.18 | 84.76 | 127.18 | 245.52 | 391.97 | 0.735 | 27.01 | 50.99 | 104.95 | 157.48 | 304.01 | 485.36 |
| 0.500 | 21.93 | 41.40 | 85.22 | 127.88 | 246.86 | 394.11 | 0.740 | 27.11 | 51.17 | 105.34 | 158.06 | 305.12 | 487.14 |
| 0.505 | 22.05 | 41.63 | 85.68 | 128.56 | 248.18 | 396.23 | 0.745 | 27.20 | 51.36 | 105.72 | 158.84 | 306.24 | 488.91 |
| 0.510 | 22.16 | 41.85 | 86.14 | 129.25 | 249.51 | 398.35 | 0.750 | 27.30 | 51.55 | 106.11 | 159.21 | 307.35 | 490.69 |
| 0.515 | 22.28 | 42.07 | 86.59 | 129.93 | 250.83 | 400.46 | 0.755 | 27.40 | 51.73 | 106.49 | 159.79 | 308.45 | 492.45 |
| 0.520 | 22.40 | 42.29 | 87.05 | 130.62 | 252.15 | 402.55 | 0.760 | 27.50 | 51.92 | 106.87 | 160.36 | 309.55 | 494.21 |
| 0.525 | 22.51 | 42.51 | 87.50 | 131.29 | 253.45 | 404.64 | 0.765 | 27.60 | 52.10 | 107.25 | 160.93 | 310.65 | 495.97 |
| 0.530 | 22.63 | 42.73 | 87.95 | 131.97 | 254.75 | 406.72 | 0.770 | 27.69 | 52.29 | 107.63 | 161.49 | 311.75 | 497.72 |
| 0.535 | 22.75 | 42.94 | 88.40 | 132.64 | 256.05 | 408.19 | 0.775 | 27.77 | 52.47 | .108.00 | 182.06 | 312.84 | 499.46 |
| 0.540 | 22.86 | 43.16 | 88.84 | 133.31 | 257.34 | 410.85 | 0.780 | 27.89 | 52.65 | 108.38 | 162.62 | 313.93 | 501.20 |
| 0.545 | 22.97 | 43.38 | 89.29 | 133.98 | 258.63 | 412.90 | 0.785 | 27.98 | 52.83 | 108.75 | 163.19 | 315.00 | 502.93 |
| 0.550 | 23.09 | 43.59 | 89.73 | 134.64 | 259.91 | 414.$4 | 0.790 | 28.08 | 53.02 | 109.13 | 163.75 | 316.10 | 504.66 |
| 0.555 | 23.20 | 43.80 | 90.17 | .135.30 | 261.18 | 416.98 | 0.795 | 28.18 | 53.20 | 109.50 | 164.31 | 317.18 | 508.39 |
| 0.560 | 23.31 | 44.02 | 90.61 | 135.96 | 262.45 | 419.01 | 0.800 | 28.27 | 53.38 | 109.87 | 164.87 | 318.26 | 508.11 |
| 0.565 | 23.43 | 44.23 | 91.04 | 136.61 | 263.71 | 421.02 | 0.805 | 28.37 | 53.56 | 110.24 | 165.42 | 319.33 | 509.82 |
| 0.570 | 23.54 | 44.44 | 91.48 | 137.26 | 264.97 | 423.03 | 0.810 | 28.46 | 53.74 | 110.61 | 165.98 | 320.40 | 511.53 |
| 0.575 | 23.65 | 44.65 | 91.91 | 137.91 | 266.23 | 425.04 | 0.815 | 28.56 | 53.92 | 110.98 | 166.53 | 321.47 | 513.23 |
| 0.580 | 23.76 | 44.86 | 92.34 | 138.56 | 267.48 | 427.03 | 0.820 | 28.65 | 54.09 | 111.35 | 167.08 | 322.54 | 514.93 |
| 0.585 | 23.87 | 45.07 | 92.77 | 139.20 | 268.72 | 429.02 | 0.825 | 28.75 | 54.27 | 111.72 | 167.63 | 323.60 | 516.63 |
| 0.590 | 23.98 | 45.28 | 93.20 | 139.85 | 269.96 | 430.99 | 0.830 | 28.84 | 54.45 | 112.08 | 168.18 | 324.66 | 518.32 |
| 0.595 | 24.09 | 45.48 | 93.62 | 140.49 | 271.19 | 482.96 | 0.835 | 28.93 | 54.63 | 112.45 | 168.73 | 325.71 | 520.00 |
| 0.600 | 24.20 | 45.69 | 94.05 | 141.12 | 272.42 | 434.93 | 0.840 | 29.03 | 54.80 | 112.81 | 169.27 | 326.76 | 521.68 |
| 0.605 | 24.31 | 45.90 | 94.97 | 141.76 | 273.65 | 436.89 | 0.845 | 29.12 | 54.98 | 113.17 | .169.82 | 327.81 | 523.36 |
| 0.610 | 24.42 | 46.10 | 94.89 | 142.39 | 27•.87 | 436:83 | 0.850 | 29.21 | 55.16 | 113.53 | 170.36 | 328.86 | 525.03 |
| 0.615 | 24.53 | 46.30 | 95.31 | 143.02 | 27′8.08 | 440.7? | 0.855 | 29.31 | 55.33 | 113.89 | 170.90 | 329.91 | 526.70 |
| 0.620 | 24.63 | 46.51 | 95.73 | 143.65 | 277.29 | 442.71 | 0.860 | 29.40 | 55.51 | 114.25 | 171.44 | 330.95 | 528.36 |
| 0.625 | 24.74 | 46.71 | 96.15 | 144.27 | 278.50 | 444.63 | 0.865 | 29.49 | 55.68 | 114.61 | 171.98 | 331.99 | 530.02 |
| 0.630 | 24.85 | 46.91 | 96.56 | 144.89 | 279.70 | 446.55 | 0.870 | 29.58 | 55.85 | 114.97 | 172.52 | 333.02 | 531.67 |
| 0.635 | 24.95 | 47.11 | 96.98 | 145.51 | 280.90 | 448.46 | 0.875 | 29.67 | 56.03 | 115.33 | 173.05 | 334.05 | 533.32 |
| 0.640 | 25.06 | 47.31 | 97.39 | 146.13 | 282.09 | 450.37 | 0.880 | 29.77 | 56.20 | 115.68 | 173.09 | 335.09 | 534.97 |
| 0.645 | 25.16 | 47.51 | 97.80 | 146.75 | 283.28 | 452.27 | 0.885 | 29.86 | 56.37 | 116.04 | 174.11 | 336.11 | 536.61 |
| 0.650 | 25.27 | 47.71 | 98.21 | 147.36 | 284.47 | 454.16 | 0.890 | 29.95 | 56.54 | 116.39 | 174.65 | 337.14 | 538.25 |
| 0.655 | 25.37 | 47.91 | 98.62 | 147.97 | 285.65 | 456.05 | 0.895 | 30.04 | 56.72 | 116.74 | 175.17 | 336.16 | 539.88 |
| 0.660 | 25.48 | 48.11 | 99.02 | 148.58 | 286.83 | 457.92 | 0.900 | 30.13 | 56.89 | 117.10 | 175.70 | 339.18 | 541.51 |
| 0.665 | 25.58 | 48.30 | 99.43 | 149.18 | 288.00 | 458.80 | 0.905 | 90.22 | 57.06 | 117.45 | 176.23 | 340.20 | 543.13 |
| 0.670 | 25.69 | 48.50 | 99.63 | 149.80 | 289.17 | 461.86 | 0.910 | 30.31 | 57.23 | 117.80 | 176.76 | 341.21 | 544.75 |
| 0.675 | 25.79 | 48.69 | 100.23 | 150.40 | 290.33 | 463.52 | 0.915 | 30.40 | 57.40 | 118.15 | 177.28 | 342.22 | 546.37 |
| 0.680 | 25.89 | 48.89 | 100.72 | 151.00 | 291.49 | 485.37 | 0.920 | 30.49 | 57.57 | 118.49 | 177.80 | 343.23 | 547.98 |
| 0.685 | 26.00 | 49.08 | 101.03 | 151.60 | 292.65 | 487.22 | 0.925 | 30.58 | 57.73 | 118.84 | 178.32 | 344.24 | 549.59 |
| 0.690 | 26.10 | 49.28 | 101.43 | 152.20 | 293.80· | 489.25 | 0.930 | 30.67 | 57.90 | 119.19 | 178.85 | 345.25 | 551.19 |
| 0.695 | 26.20 | 49.47 | 101.83 | 152.79 | 294.95 | 470.89 | 0.935 | 30.76 | 58.07 | 119.53 | 179.36 | 346.25 | 552.79 |
| 0.700 | 26.30 | 49.66 | 102.22 | 153.38 | 296.10 | 472.72 | 0.940 | 30.85 | 58.24 | 119.88 | 179.88 | 347.25 | 554.39 |
| 0.705 | 26.40. | 49.85 | 102.62 | 153.98 | 297.24 | 474.55 | 0.945 | 30.94 | 58.41 | 120.22 | 180.40 | 348.24 | 555.98 |
| 0.710 | 26.51 | 50.04 | 103.00 | 154.57 | 298.37 | 476.36 | 0.950 | 31.02 | 58.57 | 120.57 | 180.91 | 349.24 | 557.57 |
| 0.715 | 26.61 | 50.23 | 103.40 | 155.15 | 299.51 | 478.17 | 0.955 | 31.11 | 58.74 | 120.91 | 181.43 | 350.23 | 559.15 |
| 0.720 | 26.71 | 50.42 | 103.79 | 155.74 | 300.84 | 479.98 | 0.960 | 31.20 | 58.91 | 121.25 | 181.94 | 351.22 | 560.73 |

## Friction-Loss Table

| PSI PER FT. | ¾" | 1" | 1¼" | 1½" | PSI PER FT. | ¾" | 1" | PSI PER FT. | ¾" | 1" | PSI PER FT. | 1" |
|---|---|---|---|---|---|---|---|---|---|---|---|---|
| 0.965 | 31.29 | 59.07 | 121.59 | 182.45 | 1.205 | 35.28 | 66.61 | 1.445 | 38.92 | 73.48 | 1.685 | 79.84 |
| 0.970 | 31.38 | 59.24 | 121.93 | 182.96 | 1.21 | 35.36 | 66.76 | 1.450 | 38.99 | 73.62 | 1.69 | 79.97 |
| 0.975 | 31.46 | 59.4 | 122.27 | 183.47 | 1.215 | 35.44 | 66.9 | 1.455 | 39.06 | 73.75 | 1.695 | 80.1 |
| 0.980 | 31.55 | 59.57 | 122.61 | 183.98 | 1.22 | 35.52 | 67.05 | 1.460 | 39.14 | 73.88 | 1.7 | 80.22 |
| 0.985 | 31.64 | 59.73 | 122.95 | 184.49 | 1.225 | 35.59 | 67.2 | 1.465 | 39.21 | 74.03 | 1.705 | 80.35 |
| 0.990 | 31.72 | 59.89 | 123.29 | 184.99 | 1.23 | 35.67 | 67.35 | 1.470 | 39.28 | 74.16 | 1.71 | 80.48 |
| 0.995 | 31.81 | 60.06 | 123.62 | 185.5 | 1.235 | 35.75 | 67.5 | 1.475 | 39.35 | 74.3 | 1.715 | 80.61 |
| 1.000 | 31.90 | 60.22 | 123.96 | 186 | 1.24 | 35.83 | 67.65 | 1.480 | 39.42 | 74.43 | 1.72 | 80.73 |
| 1.005 | 31.98 | 60.38 | 124.29 | 186.5 | 1.245 | 35.91 | 67.79 | 1.485 | 39.5 | 74.57 | 1.725 | 80.86 |
| 1.010 | 32.07 | 60.54 | 124.63 | 187 | 1.25 | 35.98 | 67.94 | 1.490 | 39.57 | 74.71 | 1.73 | 80.99 |
| 1.015 | 32.15 | 60.71 | 124.96 | 187.5 | 1.255 | 36.06 | 68.09 | 1.495 | 39.64 | 74.84 | 1.735 | 81.11 |
| 1.020 | 32.24 | 60.87 | 125.29 | 188 | 1.26 | 36.14 | 68.23 | 1.500 | 39.71 | 74.98 | 1.74 | 81.24 |
| 1.025 | 32.32 | 61.03 | 125.62 | 188.5 | 1.265 | 36.22 | 68.38 | 1.505 | 39.78 | 75.11 | 1.745 | 81.37 |
| 1.030 | 32.41 | 61.19 | 125.95 | 189 | 1.27 | 36.29 | 68.53 | 1.510 | 39.85 | 75.25 | 1.750 | 81.49 |
| 1.035 | 32.50 | 61.35 | 126.28 | 189.49 | 1.275 | 36.37 | 68.67 | 1.515 | 39.93 | 75.38 | | |
| 1.040 | 32.58 | 61.51 | 126.61 | 189.99 | 1.28 | 36.45 | 68.82 | 1.520 | 40 | 75.52 | | |
| 1.045 | 32.66 | 61.67 | 126.94 | 190.48 | 1.285 | 36.53 | 68.96 | 1.525 | | 75.65 | | |
| 1.050 | 32.75 | 61.83 | 127.27 | 190.97 | 1.29 | 36.6 | 69.11 | 1.530 | | 75.78 | | |
| 1.055 | 32.83 | 61.99 | 127.6 | 191.46 | 1.295 | 36.68 | 69.25 | 1.535 | | 75.92 | | |
| 1.060 | 32.92 | 62.15 | 127.92 | 191.95 | 1.3 | 36.76 | 69.4 | 1.540 | | 76.05 | | |
| 1.065 | 33 | 62.31 | 128.25 | 192.44 | 1.305 | 36.83 | 69.54 | 1.545 | | 76.18 | | |
| 1.070 | 33.08 | 62.46 | 128.57 | 192.93 | 1.31 | 36.91 | 69.68 | 1.500 | | 76.32 | | |
| 1.075 | 33.17 | 62.62 | 128.9 | 193 | 1.315 | 36.98 | 69.83 | 1.555 | | 76.45 | | |
| 1.080 | 33.25 | 62.78 | 129.22 | 193.9 | 1.32 | 37.06 | 69.97 | 1.560 | | 76.58 | | |
| 1.085 | 33.33 | 62.93 | 129.55 | 194.39 | 1.325 | 37.14 | 70.11 | 1.565 | | 76.72 | | |
| 1.090 | 33.42 | 63.09 | 129.87 | 194.87 | 1.33 | 37.21 | 70.26 | 1.570 | | 76.85 | | |
| 1.095 | 33.5 | 63.25 | 130.19 | 195.35 | 1.335 | 37.29 | 70.4 | 1.575 | | 76.98 | | |
| 1.100 | 33.58 | 63.4 | 130.51 | 195.83 | 1.34 | 37.36 | 70.54 | 1.580 | | 77.11 | | |
| 1.105 | 33.66 | 63.56 | 130.83 | 196.31 | 1.345 | 37.44 | 70.68 | 1.585 | | 77.24 | | |
| 1.110 | 33.75 | 63.71 | 131.15 | 196.79 | 1.35 | 37.51 | 70.83 | 1.590 | | 77.38 | | |
| 1.115 | 33.83 | 63.87 | 131.47 | 197.27 | 1.355 | 37.59 | 70.97 | 1.595 | | 77.51 | | |
| 1.120 | 33.91 | 64.02 | 131.79 | 197.75 | 1.36 | 37.66 | 71.11 | 1.600 | | 77.64 | | |
| 1.125 | 33.99 | 64.18 | 132.11 | 198.23 | 1.365 | 37.74 | 71.25 | 1.605 | | 77.77 | | |
| 1.130 | 34.07 | 64.33 | 132.42 | 198.7 | 1.37 | 37.81 | 71.39 | 1.610 | | 77.9 | | |
| 1.135 | 34.16 | 64.49 | 132.74 | 199.18 | 1.375 | 37.89 | 71.53 | 1.615 | | 78.03 | | |
| 1.140 | 34.24 | 64.64 | 133.06 | 199.65 | 1.38 | 37.96 | 71.67 | 1.620 | | 78.16 | | |
| 1.145 | 34.32 | 64.79 | 133.37 | 200.12 | 1.385 | 38.04 | 71.81 | 1.625 | | 78.29 | | |
| 1.150 | 34.4 | 64.95 | 133.68 | 200.6 | 1.39 | 38.11 | 71.95 | 1.630 | | 78.42 | | |
| 1.155 | 34.48 | 65.1 | 134 | 201.07 | 1.395 | 38.18 | 72.09 | 1.635 | | 78.55 | | |
| 1.160 | 34.56 | 65.25 | 134.31 | 201.54 | 1.4 | 38.26 | 72.23 | 1.640 | | 78.68 | | |
| 1.165 | 34.64 | 65.4 | 134.62 | 202.01 | 1.405 | 38.33 | 72.37 | 1.645 | | 78.81 | | |
| 1.170 | 34.72 | 65.55 | 134.94 | 202.47 | 1.41 | 38.4 | 72.51 | 1.650 | | 78.94 | | |
| 1.175 | 34.8 | 65.71 | 135.25 | 202.94 | 1.415 | 38.48 | 72.65 | 1.655 | | 79.07 | | |
| 1.180 | 34.88 | 65.86 | 135.56 | 203.41 | 1.42 | 38.55 | 72.79 | 1.680 | | 79.2 | | |
| 1.185 | 34.96 | 66.01 | 135.87 | 203.87 | 1.425 | 38.63 | 72.93 | 1.665 | | 79.33 | | |
| 1.190 | 35.04 | 66.16 | 136.18 | 204.34 | 1.43 | 38.7 | 73.06 | 1.670 | | 79.46 | | |
| 1.195 | 35.12 | 66.31 | 136.49 | 204.8 | 1.435 | 38.77 | 73.2 | 1.675 | | 79.58 | | |
| 1.200 | 35.2 | 66.46 | 136.8 | 205.26 | 1.44 | 38.85 | 73.34 | 1.680 | | 79.71 | | |

# Friction-Loss Table

| PSI PER FT. | 3″ | 3½″ | 4″ | PSI PER FT. | :r | 3″ | PSI PER FT. | 5″ | 8″ | 8″ | PSI PER FT. | 6″ |
|---|---|---|---|---|---|---|---|---|---|---|---|---|
| 0.005 | 57.92 | 84.93 | 118.45 | 0.245 | 474.77 | 696.09 | 0.0005 | 61.9 | 100.3 | 212.9 | 0.0285 | 892.4 |
| 0.010 | 84.25 | 123.53 | 172.29 | 0.250 | 479.98 | 703.73 | 0.0010 | 90.0 | 145.9 | 309.6 | 0.0290 | 900.9 |
| 0.015 | 104.90 | 153.80 | 214.51 | 0.255 | 485.15 | 711.30 | 0.0015 | 112.0 | 181.7 | 385.5 | 0.0295 | 909.2 |
| 0.020 | 122.55 | 179.67 | 250.60 | 0.260 | 490.27 | 718.81 | 0.0020 | 130.9 | 212.3 | 450.4 | 0.0300 | 917.5 |
| 0.025 | 138.26 | 202.71 | 282.72 | 0.265 | 495.34 | 726.25 | 0.0025 | 147.6 | 239.5 | 508.1 | 0.0305 | 925.8 |
| 0.030 | 152.58 | 223.70 | 312.00 | 0.270 | 500.95 | 733.62 | 0.0030 | 162.9 | 264.3 | 560.7 | 0.0310 | 933.9 |
| 0.035 | 155.83 | 243.14 | 339.12 | 0.275 | 505.36 | 740.94 | 0.0035 | 177.1 | 287.3 | 609.4 | 0.0315 | 942.1 |
| 0.040 | 178.25 | 261.34 | 364.50 | 0.280 | 510.31 | 748.19 | 0.0040 | 190.4 | 308.8 | 655.1 | 0.0320 | 950.1 |
| 0.045 | 189.96 | 278.52 | 388.46 | 0.285 | 515.21 | 755.38 | 0.0045 | 202.9 | 329.0 | 698.1 | 0.0325 | 958.1 |
| 0.050 | 201.10 | 294.84 | 411.23 | 0.290 | 520.08 | 762.52 | 0.0050 | 214.8 | 348.3. | 739.1 | 0.0330 | 966.0 |
| 0.055 | 211.73 | 310.43 | 432.97 | 0.295 | 524.91 | 769.59 | 0.0055 | 226.1 | 366.8 | 778.1 | 0.0335 | 973.9 |
| 0.060 | 221.92 | 325.38 | 453.82 | 0.300 | 529.70 | 776.62 | 0.0060 | 237.0 | 384.4 | 815.6 | 0.0340 | 981.8 |
| 0.065 | 231.74 | 339.76 | 473.88 | 0.305 | 534.45 | | 0.0065 | 247.5 | 401.4 | 851.7 | 0.0345 | 989.5 |
| 0.070 | 241.21 | 353.65 | 493.25 | 0.310 | 539.17 | | 0.0070 | 257.6 | 417.8 | 886.5 | 0.0350 | 997.3 |
| 0.075 | 250.37 | 367.09 | 511.99 | 0.315 | 543.85 | | 0.0075 | 267.4 | 433.7 | 920.2 | 0.0355 | 1004.9 |
| 0.080 | 259.26 | 380.12 | 530.17 | 0.320 | 548.50 | | 0.0080 | 276.9 | 449.1 | 952.8 | 0.0360 | 1012.6 |
| 0.085 | 267.90 | 392.78 | 547.83 | 0.325 | 553.12 | | 0.0085 | 286.1 | 464.1 | 984.6 | 0.0365 | 1020. |
| 0.090 | 276.31 | 405.11 | 565.02 | 0.330 | 557.70 | | 0.0090 | 295.1 | 478.6 | 1015.5 | 0.0370 | 1027.7 |
| 0.095 | 284.50 | 417.12 | 581.78 | 0.335 | 562.25 | | 0.0095 | 303.8 | 492.8 | 1045.6 | 0.0375 | 1035.2 |
| 0.100 | 292.50 | 428.85 | 598.18 | 0.340 | 566.77 | | 0.0100 | 312.4 | 506.7 | 1075.0 | 0.0380 | 1042.6 |
| 0.105 | 300.31 | 440.31 | 614.12 | 0.345 | 571.26 | | 0.0105 | 320.7 | 520.2 | 1103.7 | 0.0385 | 1050.0 |
| 0.110 | 307.96 | 451.52 | 629.76 | 0.350 | 575.72 | | 0.0110 | 328.9 | 533.4 | 1131.8 | 0.0390 | 1057.3 |
| 0.115 | 315.45 | 462.50 | 645.07 | 0.355 | 580.15 | | 0.0115 | 336.9 | 546.4 | 1159.3 | 0.0395 | 1064.6 |
| 0.120 | 322.79 | 473.27 | 660.08 | 0.360 | 584.56 | | 0.0120 | 344.7 | 559.1 | 1186.3 | 0.0400 | 1071.9 |
| 0.125 | 329.99 | 483.82 | 674.81 | 0.365 | 588.93 | | 0.0125 | 352.4 | 571.6 | 1212.8 | | |
| 0.130 | 337.07 | 494.19 | 689.27 | 0.370 | 593.28 | | 0.0130 | 360.0 | 583.9 | 1238.5 | | |
| 0.135 | 344.01 | 504.38 | 703.48 | 0.375 | 597.60 | | 0.0135 | 367.4 | 595.9 | 1264.3 | | |
| 0.140 | 350.84 | 514.39 | 717.44 | 0.380 | 601.89 | | 0.0140 | 374.7 | 607.7 | 1289.4 | | |
| 0.145 | 357.56 | 524.24 | 731.18 | 0.385 | 606.16 | | 0.0145 | 381.8 | 619.4 | 1314.1 | | |
| 0.150 | 364.17 | 533.93 | 744.71 | 0.390 | 610.40 | | 0.0150 | 388.9 | 630.8 | 1338.4 | | |
| 0.155 | 370.69 | 543.48 | 758.02 | 0.395 | 614.62 | | 0.0155 | 395.9 | 642.1 | 1362.3 | | |
| 0.160 | 377.10 | 552.89 | 771.15 | 0.400 | 618.81 | | 0.0160 | 402.7 | 653.2 | 1385.9 | | |
| 0.165 | 383.43 | 562.16 | 784.08 | 0.405 | 622.98 | | 0.0165 | 409.5 | 664.2 | 1409.1 | | |
| 0.170 | 389.66 | 571.31 | 796.83. | 0.410 | 627.13 | | 0.0170 | 416.1 | 675.0 | 1432.1 | | |
| 0.175 | 395.82 | 580.33 | 809.42 | 0.415 | 631.25 | | 0.0175 | 422.7 | 685.6 | 1454.7 | | |
| 0.180 | 401.89 | 589.24 | 821.84 | 0.420 | 635.35 | | 0.0180 | 429.2 | 696.1 | 1477.0 | | |
| 0.185 | 407.89 | 598.03 | 834.10 | 0.425 | 639.43 | | 0.0185 | 435.6 | 706.5 | 1499.0 | | |
| 0.190 | 413.81 | 606.71 | 846.21 | 0.430 | 643.48 | | 0.0190 | 441.9 | 716.8 | 1520.8 | | |
| 0.195 | 419.66 | 615.29 | 858.18 | 0.435 | 647.52 | | 0.0195 | 448.2 | 726.9 | 1542.3 | | |
| 0.200 | 425.44 | 623.77 | 870.00 | 0.440 | 651.53 | | 0.0200 | 454.3 | 736.9 | 1563.6 | | |
| 0.205 | 431.16 | 632.15 | 881.69 | 0.445 | 655.52 | | 0.0205 | 461.0 | 746.8 | 1584.6 | | |
| 0.210 | 436.81 | 640.44 | 893.25 | 0.450 | 659.49 | | 0.0210 | 466.5 | 756.6 | 1605.3 | | |
| 0.215 | 442.40 | 648.63 | 904.69 | 0.455 | 663.44 | | 0.0215 | 472.4 | 766.3 | 1625.9 | | |
| 0.220 | 447.94 | 656.75 | 916.00 | 0.460 | 667.38 | | 0.0220 | 478.4 | 775.9 | 1646.2 | | |
| 0.225 | 453.41 | 664.77 | 927.19 | 0.465 | 671.29 | | 0.0225 | 484.2 | 785.4 | 1666.3 | | |
| 0.230 | 458.83 | 672.72 | 938.27 | 0.470 | 675.18 | | 0.0230 | 490.0 | 794.8 | 1686.2 | | |
| 0.235 | 464.19 | 680.58 | 949.25 | 0.475 | 679.05 | | 0.0235 | 495.7 | 804.1 | 1706.0 | | |
| 0.240 | 469.51 | 688.37 | 960.11 | 0.480 | 682.91 | | 0.0240 | 501.4 | 813.3 | 1725.5 | | |
| | | | | 0.48!5 | 686.74 | | 0.0245 | 507.0 | 822.4 | 1744.8 | | |
| | | | | 0.490 | 690.56 | | 0.0250 | 512.6 | 831.4 | 1764.0 | | |
| | | | | 0.495 | 694.36 | | 0.0255 | 518.1 | 840.4 | 1783.0 | | |

(continued)

(continued)

| PSI PER FT. | 3″ | 3½″ | 4″ | PSI PER FT. | :r | 3″ | PSI PER FT. | 5″ | 8″ | 8″ | PSI PER FT. | 6″ |
|---|---|---|---|---|---|---|---|---|---|---|---|---|
| | | | | 0.500 | 698.14 | | 0.0260 | 523.6 | 849.2 | 1801.8 | | |
| | | | | | | | 0.0265 | 529.0 | 858.0 | 1820.4 | | |
| | | | | | | | 0.0270 | 534.4 | 866.7 | 1838.9 | | |
| | | | | | | | 0.0275 | 539.7 | 875.4 | 1857.2 | | |
| | | | | | | | 0.0280 | 545.6 | 883.9 | 1875.4 | | |

# Appendix F
# Pipe Schedule System-Past and Present

| Pipe size inches | 1953 Schedule | | | 1940 Schedule | | | | | |
|---|---|---|---|---|---|---|---|---|---|
| | Hazard of occupancy | | Hazard of occupancy | | | | 1905 1–2–3 Schedule | 1896 1–2–4 Schedule | Pre-1896 1–3–6 Schedule |
| | Light | Ordinary | Extra | Light | Ordinary | Extra | | | |
| ¾ | 0 | 0 | 0 | | Same as 1953 Schedule | | 1 | 1 | 1 |
| 1 | 2 | 2 | 1 | | | | 2 | 2 | 3 |
| 1¼ | 3 | 3 | 2 | | | | 3 | 4 | 6 |
| 1½ | 5 | 5 | 5 | | | | 5 | 8 | 10 |
| 2 | 10 | 10 | 8 | | | | 10 | 16 | 18 |
| 2½ | 30 | 20 | 15 | 40 | 20 | 15 | 20 | 28 | 28 |
| 3 | 60 | 40 | 27 | No limit | 40 | 27 | 36 | 48 | 48 |
| 3½ | 100 | 65 | 40 | – | 65 | 40 | 55 | 78 | 78 |
| 4 | No limit | 100 | 55 | – | 100 | 55 | 80 | 110 or 115 | 115 |
| 5 | – | 160 | 90 | – | 160 | 90 | 140 | 150 | 200 |
| 6 | – | 275 | 150 | – | 250 | 150 | 200 | 200 | 350 |
| 8 | – | 400 | – | – | – | – | – | – | – |

The 1905 Schedule was the Mutual Schedule starting in 1895, but not adopted by the stock companies until 1905.

Starting in 1955, for ordinary hazard schedules, if the distance between heads exceeds 12 feet or the distance between branch lines exceeds 12 feet, lower limits apply to the number of heads supplied by certain pipe sizes as follows: 2½″, 15; 3″, 30; 3½″, 60.

Under the current ordinary hazard schedule, if there are 9 heads on a branch line, the second piece of pipe from the end must be increased from 1″ to 1¼″ .The same applies to a 10-head branch line and, in addition, the pipe feeding the tenth sprinkler must be 2½″.

A separate schedule is provided for copper tubing. Refer to NFPA 13.



H. S. Wass Jr. and R. P. Fleming P.E., *Sprinkler Hydraulics*,
https://doi.org/10.1007/978-3-030-02595-3

# Appendix G

Data on some currently listed sprinkler pipe.

ID: Internal diameter in inches.

OD: Outside diameter in inches.

FLC: Friction loss constant—See Table A.4 of Appendix A.

VPC: Velocity pressure constant—See Table A.5 of Appendix A.

*Note* The friction-loss table in Appendix E may be used if you apply the modification factor shown in that Appendix.

Allied "XL"

| Nom. pipe size | ID | OD | C = 120 FLC | C = 100 FLC | VPC |
|---|---|---|---|---|---|
| 1 | 1.I25 | 1.295 | $3.627 \times 10^{-5}$ | $5.082 \times 10^{-4}$ | $7.01 \times 10^{-5}$ |
| 1¼ | 1.475 | 1.645 | $9.697 \times 10^{-5}$ | $1.359 \times 10^{-5}$ | $2.37 \times 10^{-4}$ |
| 1½ | 1.71 0 | 1.890 | $4.720 \times 10^{-5}$ | $6.614 \times 10^{-5}$ | $1.31 \times 10^{-4}$ |
| 2 | 2.177 | 2.367 | $1.456 \times 10^{-5}$ | $2.041 \times 10^{-5}$ | $5.00 \times 10^{-5}$ |
| 2 | 2.607 | 2.867 | $6.054 \times 10^{-6}$ | $8.483 \times 10^{-6}$ | $2.43 \times 10^{-5}$ |
| 3 | 3.226 | 3.486 | $2.145 \times 10^{-6}$ | $3.006 \times 10^{-6}$ | $1.04 \times 10^{-5}$ |

"POZ-LOCK"

| Nom. pipe size | ID | OD | C = 120 FLC | C = 100 FLC | VPC |
|---|---|---|---|---|---|
| 1 | 1.004 | 1.080 | $6.313 \times 10^{-5}$ | $8.845 \times 1\,0^{-4}$ | $1.11 \times 10^{-3}$ |
| 1¼ | 1.227 | 1.315 | $2.377 \times 10^{-5}$ | $3.330 \times 10^{-4}$ | $4.95 \times 10^{-4}$ |
| 1½ | 1.560 | 1.660 | $7.387 \times 10^{-5}$ | $1.034 \times 10^{-4}$ | $1.90 \times 10^{-4}$ |



H. S. Wass Jr. and R. P. Fleming P.E., *Sprinkler Hydraulics*,
https://doi.org/10.1007/978-3-030-02595-3

Central Sprinkler "TL" Threadable Lightwall

| Nom. pipe size | ID | OD | C = 120 FLC | C = 100 FLC | VPC |
|---|---|---|---|---|---|
| 1 | 1.110 | 1.290 | $3.872 \times 10^{-4}$ | $5.425 \times 10^{-4}$ | $7.40 \times 10^{-4}$ |
| 1¼ | 1.458 | 1.638 | $1.026 \times 10^{-4}$ | $1.438 \times 10^{-4}$ | $2.49 \times 10^{-4}$ |
| 1½ | 1.693 | 1.883 | $4.956 \times 10^{-5}$ | $6.944 \times 10^{-5}$ | $1.37 \times 10^{-4}$ |
| 2 | 2.160 | 2.360 | $1.513 \times 10^{-5}$ | $2.120 \times 10^{-5}$ | $5.16 \times 10^{-5}$ |

American Tube Dyna Thread-40

| Nom. pipe size | ID | OD | C = 120 FLC | C = 100 FLC | VPC |
|---|---|---|---|---|---|
| 1 | 1.087 | 1.315 | $4.288 \times 10^{-4}$ | $6.008 \times 10^{-4}$ | $8.04 \times 10^{-4}$ |
| 1¼ | 1.416 | 1.660 | $1.183 \times 10^{-4}$ | $1.658 \times 10^{-4}$ | $2.79 \times 10^{-4}$ |
| 1½ | 1.650 | 1.900 | $5.617 \times 10^{-5}$ | $7.870 \times 10^{-5}$ | $1.52 \times 10^{-4}$ |
| 2 | 2.115 | 2.375 | $1.676 \times 10^{-5}$ | $2.349 \times 10^{-5}$ | $5.61 \times 10^{-5}$ |

American Tube Black Lightwall Threadable (BLT)

| Nom. pipe size | ID | OD | C = 120 FLC | C = 100 FLC | VPC |
|---|---|---|---|---|---|
| 1 | 1.104 | 1.290 | $3.976 \times 10^{-4}$ | $5.570 \times 10^{-4}$ | $7.56 \times 10^{-4}$ |
| 1¼ | 1.452 | 1.638 | $1.047 \times 10^{-4}$ | $1.467 \times 10^{-4}$ | $2.53 \times 10^{-4}$ |
| 1½ | 1.687 | 1.883 | $5.042 \times 10^{-5}$ | $7.065 \times 10^{-5}$ | $1.39 \times 10^{-4}$ |
| 2 | 2.154 | 2.360 | $1.534 \times 10^{-5}$ | $2.149 \times 10^{5}$ | $5.22 \times 10^{-5}$ |

American Tube Dyna Flow-I0

| Nom. pipe size | ID | OD | C = 120 FLC | C = 100 FLC | VPC |
|---|---|---|---|---|---|
| 1 | 1.197 | 1.315 | $2.681 \times 10^{-4}$ | $3.757 \times 10^{-4}$ | $5.47 \times 10^{-4}$ |
| 1¼ | 1.542 | 1.660 | $7.811 \times 10^{-5}$ | $1.094 \times 10^{-4}$ | $1.99 \times 10^{-4}$ |
|  | 1.752 | 1.900 | $4.194 \times 10^{-5}$ | $5.877 \times 10^{-5}$ | $1.19 \times 10^{-4}$ |
| 2 | 2.227 | 2.375 | $1.304 \times 10^{-5}$ | $1.827 \times 10^{-5}$ | $4.57 \times 10^{-5}$ |
| 2½ | 2.731 | 2.875 | $4.828 \times 10^{-6}$ | $6.765 \times 10^{-6}$ | $2.02 \times 10^{-5}$ |
| 3 | 3.340 | 3.500 | $1.811 \times 10^{-6}$ | $2.538 \times 10^{-6}$ | $9.02 \times 10^{-6}$ |
| 4 | 4.328 | 4.500 | $5.128 \times 10^{-7}$ | $7.185 \times 10^{-7}$ | $3.20 \times 10^{-6}$ |

American Tube Dyna Light-S

| Nom. pipe size | ID | C = 120<br>OD | C = 100<br>FLC | FLC | VPC |
|---|---|---|---|---|---|
| 1 | 1.185 | 1.315 | $2.816 \times 10^{-4}$ | $3.946 \times 10^{-4}$ | $5.70 \times 10^{-4}$ |
| 1¼ | 1.530 | 1.660 | $8.113 \times 10^{-4}$ | $1.137 \times 10^{-4}$ | $2.05 \times 10^{-4}$ |
| 1½ | 1.770 | 1.900 | $3.990 \times 10^{-5}$ | $5.591 \times 10^{-5}$ | $1.14 \times 10^{-4}$ |
| 2 | 2.245 | 2.375 | $1.254 \times 10^{-5}$ | $1.757 \times 10^{-5}$ | $4.42 \times 10^{-5}$ |

Central Sprinkler "Schedule 7"

| Nom. pipe size | ID | OD | C = 120<br>FLC | C = 100<br>FLC | VPC |
|---|---|---|---|---|---|
| 1½ | 1.752 | 1.900 | $4.194 \times 10^{-5}$ | $5.877 \times 10^{-5}$ | $1.19 \times 10^{-4}$ |
| 2 | 2.227 | 2.375 | $1.304 \times 10^{-5}$ | $1.827 \times 10^{-5}$ | $4.57 \times 10^{-5}$ |
| 2½ | 2.695 | 2.875 | $5.750 \times 10^{-6}$ | $7.216 \times 10^{-6}$ | $2.13 \times 10^{-5}$ |
| 3 | 3.320 | 3.500 | $1.865 \times 10^{-6}$ | $2.613 \times 10^{-6}$ | $9.24 \times 10^{-6}$ |
| 4 | 4.300 | 4.400 | $5.293 \times 10^{-7}$ | $7.416 \times 10^{-7}$ | $3.29 \times 10^{-6}$ |

Polybutylene plastic pipe: manufactured in two sets of nominal sizes, "copper tube size" (CTS) and "iron pipe size" (IPS).

Polybutylene—CTS

| Nom. pipe size | ID | OD | FLC | VPC |
|---|---|---|---|---|
| 1 | 0.911 | 1.125 | $6.707 \times 10^{-4}$ | $1.63 \times 10^{-3}$ |
| 1¼ | 1.112 | 1.375 | $2.540 \times 10^{-4}$ | $7.34 \times 10^{-4}$ |
| 1½ | 1.314 | 1.625 | $1.127 \times 10^{-4}$ | $3.77 \times 10^{-4}$ |
| 2 | 1.720 | 2.125 | $3.306 \times 10^{-5}$ | $1.28 \times 10^{-4}$ |

Polybutylene—IPS

| Nom. pipe size | ID | OD | FLC | VPC |
|---|---|---|---|---|
| 1 | 1.051 | 1.315 | $3.343 \times 10^{-4}$ | $9.20 \times 10^{-4}$ |
| 1¼ | 1.332 | 1.660 | $1.054 \times 10^{-4}$ | $3.57 \times 10^{-4}$ |
| 1½ | 1.528 | 1.900 | $5.404 \times 10^{-5}$ | $2.06 \times 10^{-4}$ |
| 2 | 1.917 | 2.375 | $1.791 \times 10^{-5}$ | $8.32 \times 10^{-5}$ |

Post-chlorinated polyvinyl chloride (CPVC) plastic pipe.

| Nom. pipe size | ID | OD | FLC | VPC |
|---|---|---|---|---|
| 1 | 1.1 09 | 1.315 | $2.574 \times 10^{-4}$ | $7.42 \times 10^{-4}$ |
| 1¼ | 1.400 | 1.660 | $8.274 \times 10^{-5}$ | $2.92 \times 10^{-4}$ |
| 1½ | 1.602 | 1.900 | $4.292 \times 10^{-5}$ | $1.71 \times 10^{-4}$ |
| 2 | 2.003 | 2.375 | $1.446 \times 10^{-5}$ | $6.98 \times 10^{-5}$ |
| 2½ | 2.423 | 2.875 | $5.722 \times 10^{-6}$ | $3.26 \times 10^{-5}$ |
| 3 | 2.951 | 3.500 | $2.191 \times 10^{-6}$ | $1.48 \times 10^{-5}$ |

Type L copper

| Nom. pipe size % | ID | FLC | VPC |
|---|---|---|---|
| ¾ | 0.785 | $2.092 \times 10^{-3}$ | $2.96 \times 10^{-3}$ |
| 1 | 1.025 | $5.707 \times 10^{-4}$ | $1.02 \times 10^{-3}$ |
| 1¼ | 1.265 | $2.049 \times 10^{-4}$ | $4.39 \times 10^{-4}$ |
| 1½ | 1.505 | $8.791 \times 10^{-5}$ | $2.19 \times 10^{-5}$ |
| 2 | 1.985 | $2.283 \times 10^{-5}$ | $7.23 \times 10^{-5}$ |
| 2½ | 2.465 | $7.953 \times 10^{-6}$ | $3.04 \times 10^{-5}$ |
| 3 | 2.945 | $3.344 \times 10^{-6}$ | $1.49 \times 10^{-5}$ |
| 3½ | 3.425 | $1.603 \times 10^{-6}$ | $8.16 \times 10^{-6}$ |
| 4 | 3.905 | $8.462 \times 10^{-7}$ | $4.83 \times 10^{-6}$ |
| 5 | 4.875 | $2.872 \times 10^{-7}$ | $1.99 \times 10^{-6}$ |
| 6 | 5.845 | $1.187 \times 10^{-7}$ | $9.62 \times 10^{-7}$ |
| 8 | 7.725 | $3.052 \times 10^{-8}$ | $3.15 \times 10^{-7}$ |
| 10 | 9.625 | $1.046 \times 10^{-8}$ | $1.31 \times 10^{-7}$ |
| 12 | 11.565 | $4.277 \times 10^{-9}$ | $6.28 \times 10^{-8}$ |

# Index




H. S. Wass Jr. and R. P. Fleming P.E., *Sprinkler Hydraulics*,
https://doi.org/10.1007/978-3-030-02595-3

MIX
Papier aus verantwortungsvollen Quellen
Paper from responsible sources
FSC® C105338

If you have any concerns about our products,
you can contact us on
**ProductSafety@springernature.com**

In case Publisher is established outside the EU,
the EU authorized representative is:
**Springer Nature Customer Service Center GmbH**
**Europaplatz 3, 69115 Heidelberg, Germany**

Printed by Libri Plureos GmbH
in Hamburg, Germany